Arduino®

2nd Edition

by John Nussey

Arduino® For Dummies®, 2nd Edition

Published by: **John Wiley & Sons, Inc.**, 111 River Street, Hoboken, NJ 07030-5774, www.wiley.com

For general information on our other products and services, please contact our Customer Care Department within the U.S. at 877-762-2974, outside the U.S. at 317-572-3993, or fax 317-572-4002. For technical support, please visit https://hub.wiley.com/community/support/dummies.

Wiley publishes in a variety of print and electronic formats and by print-on-demand. Some material included with standard print versions of this book may not be included in e-books or in print-on-demand. If this book refers to media such as a CD or DVD that is not included in the version you purchased, you may download this material at http://booksupport.wiley.com. For more information about Wiley products, visit www.wiley.com.

Library of Congress Control Number: 2018951004

ISBN 978-1-119-48954-2 (pbk); ISBN 978-1-119-48955-9 (ebk); ISBN 978-1-119-48957-3 (ebk)

Manufactured in the United States of America

C10003211_080918

Arduino®

2nd Edition

by John Nussey

Arduino® For Dummies®, 2nd Edition

Published by: **John Wiley & Sons, Inc.**, 111 River Street, Hoboken, NJ 07030-5774, www.wiley.com

Published simultaneously in Canada

For general information on our other products and services, please contact our Customer Care Department within the U.S. at 877-762-2974, outside the U.S. at 317-572-3993, or fax 317-572-4002. For technical support, please visit https://hub.wiley.com/community/support/dummies.

Wiley publishes in a variety of print and electronic formats and by print-on-demand. Some material included with standard print versions of this book may not be included in e-books or in print-on-demand. If this book refers to media such as a CD or DVD that is not included in the version you purchased, you may download this material at http://booksupport.wiley.com. For more information about Wiley products, visit www.wiley.com.

Library of Congress Control Number: 2018951004

ISBN 978-1-119-48954-2 (pbk); ISBN 978-1-119-48955-9 (ebk); ISBN 978-1-119-48957-3 (ebk)

Manufactured in the United States of America

C10003211_080918

Contents at a Glance

Table of Contents

Foreword

The moment a *For Dummies* book comes out, it's definitely a milestone in the history of a product.

Programming embedded computers used to be a very difficult task, reserved only to experienced engineers willing to master the obscure assembly language. In recent years, however, many platforms have tried to make this task simpler and more accessible to everyday people. Arduino is one of the latest attempts at making technology less scary and more creative.

With John, this book's author, we watched this creative tool being adopted by designers and artists in London, making its way into many memorable projects. Now Arduino has escaped the lab of Arts & Design and spread like a virus, becoming the tool of choice for all kinds of people who have great ideas they want to realize.

I'm really glad that John decided to write this book, because he's an early user of the Arduino platform from back in the days when it was still quite experimental. Having taught Arduino classes for many years, he has the ability to introduce the subject to all audiences.

Any newcomer to Arduino will, with the right tools and teaching — such as those found in this book — show true genius in no time.

— Massimo Banzi

Introduction

Arduino is a tool, a community, and a way of thinking that is affecting how we use and understand technology. It has rekindled a love and understanding of electronics for many people, including myself, who felt that electronics was something that they had left behind at school.

Arduino is tiny circuit board that has huge potential. It can be used to blink a Morse-code signal using a single light-emitting diode (LED) or to control every light in a building, depending on how far you take it. Its capabilities are limited only by your imagination.

Arduino is also providing a new, practical approach to technical education, lowering the entry level for those wanting to use electronics to complete small projects and, I hope, encouraging you to read further to take on big ones.

A huge and ever-growing community of Arduinists has emerged — users and developers who learn from each other and contribute to the open-source philosophy by sharing the details of their projects. This open-source attitude is responsible for the huge popularity of Arduino.

Arduino is more than just a gadget; it's a tool. A piece of technology that makes understanding and using today's technology easier.

So if the prospect of understanding the limitless possibilities of technology doesn't sound interesting to you, please put this book down and back away.

Otherwise, read on!

About This Book

This is a technical book, but it's not for technical people only. Arduino is designed to be usable by anyone, whether they're technical, creative, crafty, or just curious. All you need is an open mind or a problem to fix and you'll soon find ways that using Arduino can benefit you.

This book starts on the most basic level to get you started with using and understanding Arduino. At times throughout the book, I may refer to a number of technical things that will, like anything, take time to understand. I guide you through all the basics and then on to more advanced activities.

Much of what is in this book is based on my learning and teaching experiences. I learned all about Arduino from scratch, but have always found that the best way to learn is in practice, by making your own projects. The key is to understand the basics that I cover in this book and then build on that knowledge by thinking about how you can apply it to solve problems, create things, or just entertain yourself.

Foolish Assumptions

I assume nothing about your technical knowledge. Arduino is an easy-to-use platform for learning about electronics and programming. It is for people from all walks of life, whether you're a designer, an artist, or a hobbyist.

It can also be a great platform for people who are already technical. Maybe you've done a bit of coding but want to bring your projects into the physical world in some way. Or maybe you've worked with electronics and want to see what Arduino can bring to the table.

Whoever you are, you'll find that Arduino has great potential. It's really up to you to decide what to make of it.

Icons Used in This Book

Arduino For Dummies uses icons to highlight important points for you. Keep an eye out for the following:

TIP

This icon highlights a bit of helpful information. That info may be a technique to help you complete a project more easily or the answer to a common problem.

WARNING

Arduinos aren't dangerous on their own; indeed, they're made to be extremely safe and easy to use. But if you use them in a circuit without proper planning as well as care and attention, they can damage your circuit, your computer, and yourself. When you see a Warning icon, please take special note.

Often, you must consider certain points before proceeding with a task. I use Remember icons to remind you of such points.

Some information is more technical than others and is not for the faint-hearted. The joy of Arduino is that you don't need to fully understand the technical details immediately. You can skip anything that's marked with this icon if it's more complicated than you want to deal with at the moment; you can always return to it when you're ready.

Beyond the Book

In addition to what you're reading right now, this product comes with a free access-anywhere Cheat Sheet that provides information on using resistors, getting the tools you'll need, and some system shortcuts. To get this Cheat Sheet, simply go to `www.dummies.com` and type *Arduino For Dummies 2nd Edition Cheat Sheet* in the Search box. I also provide a bonus chapter that teaches you all about using your Arduino to hack other hardware, such as games, controllers, and toys.

Where to Go from Here

If you're uncertain about where to start, I suggest the beginning. By the end of Chapter 2, you'll have acquired a simple understanding of Arduino and will know where you can get a kit to continue learning.

If you've used Arduino before, you may want to jump straight to Chapter 4 to cover the basics again, or head straight to the area that interests you.

1

Getting to Know Arduino

IN THIS PART . . .

Find out all about the little blue circuit board.

Discover everything you need to get started with Arduino and where to get them.

Learn how to wield the awesome power of an LED, blinking in on command with a few simple lines of code.

» Learning where Arduino came from

» Introducing the basic principles

Chapter 1
Discovering Arduino

Arduino is made up of both hardware and software.

The Arduino board is a printed circuit board (PCB) designed to use a microcontroller chip as well as other input and outputs. The board has many other electronic components that are needed for the microcontroller to function or to extend its capabilities.

A microcontroller is a small computer contained in a single, integrated circuit or computer chip. Microcontrollers are an excellent way to program and control electronics. *Microcontroller boards* have a microcontroller chip and other useful connectors and components that allow a user to attach inputs and outputs. Some examples of devices with microcontroller boards are the Wiring board, the PIC, and the Basic Stamp.

You write code in the Arduino software to tell the microcontroller what to to-do. For example, by writing a line of code, you can tell an light-emitting diode (LED) to blink on and off. If you connect a pushbutton and add another line of code, you can tell the LED to turn on only when the button is pressed. Next, you may want to tell the LED to blink only when the pushbutton is held down. In this way, you can quickly build a behavior for a system that would be difficult to achieve without a microcontroller.

Similar to a conventional computer, an Arduino can perform a multitude of functions, but it's not much use on its own. It requires inputs or outputs to make it useful. These inputs and outputs allow a computer — and an Arduino — to sense objects in the world and to affect the world.

Before you move forward, it might help you to understand a bit of the history of Arduino.

Where Did Arduino Come From?

Arduino started its life in Italy, at Interaction Design Institute Ivrea (IDII), a graduate school for interaction design that focuses on how people interact with digital products, systems, and environments and how they in turn influence us.

The term *interaction design* was coined by Bill Verplank and Bill Moggridge in the mid-1980s. The sketch in Figure 1-1 by Verplank illustrates the basic premise of interaction design: If you do something, you feel a change, and from that you can know something about the world.

FIGURE 1-1: The principle of interaction design, illustrated by Bill Verplank.

Courtesy of Bill Verplank

Although interaction design is a general principle, it more commonly refers specifically to how we interact with conventional computers by using peripherals (such as mice, keyboards, and touchscreens) to navigate a digital environment that is graphically displayed on a screen.

Another avenue, referred to as *physical computing*, is about extending the range of these computer programs, software, or systems through electronics. By using electronics, computers can sense more about the world and have a physical effect on the world themselves.

Both areas — interaction design and physical computing — require prototypes to fully understand and explore the interactions, which presented a hurdle for non-technical design students.

In 2001, a project called Processing, started by Casey Reas and Benjamin Fry, aimed to get non-programmers into programming by making it quick and easy to produce onscreen visualizations and graphics. The project gave the user a digital sketchbook on which to try ideas and experiment with a small investment of time. This project in turn inspired a similar project for experimenting in the physical world.

In 2003, building on the same principles as Processing, Hernando Barragán started developing a microcontroller board called Wiring. This board was the predecessor to Arduino.

In common with the Processing project, the Wiring project also aimed to involve artists, designers, and other non-technical people. However, Wiring was designed to get people into electronics as well as programming. The Wiring board (shown in Figure 1-2) was less expensive than some other microcontrollers, such as the PIC and the Basic Stamp, but it was still a sizable investment for students.

In 2005, the Arduino project began in response to the need for affordable and easy-to-use devices for interaction design students to use in their projects. It is said that Massimo Banzi and David Cuartielles named the project after Arduin of Ivrea, an Italian king, but I've heard from reliable sources that it also happens to be the name of the local pub near the university, which may have been of more significance to the project.

The Arduino project drew from many of the experiences of both Wiring and Processing. For example, an obvious influence from Processing is the *graphic user interface* (GUI) in the Arduino software. This GUI was initially "borrowed" from Processing, and even though it still looks similar, it has since been refined to be more specific to Arduino. I cover the Arduino interface in more depth in Chapter 3.

FIGURE 1-2: An early Wiring board.

Arduino also kept the naming convention from Processing, calling its programs *sketches.* In the same way that Processing gives people a digital sketchbook to create and test programs quickly, Arduino gives people a way to sketch their hardware ideas as well. Throughout this book, I show many sketches that allow your Arduino to perform a huge variety of tasks. By using and editing the example sketches in this book, you can quickly build up your understanding of how they work. You'll be writing your own in no time. Each sketch is followed with a line-by-line explanation of how it works to ensure that no stone is left unturned.

The Arduino board, shown in Figure 1-3, was made to be more robust and forgiving than Wiring and other earlier microcontrollers. It was not uncommon for students, especially those from a design or arts background, to break their microcontroller within minutes of using it, simply by getting the wires the wrong way around. This fragility was a huge problem, not only financially but also for the success of the boards outside technical circles. You can also change the microcontroller chip on an Arduino; if the chip becomes damaged, you can replace just it rather than the entire board.

Another important difference between Arduino and other microcontroller boards is the cost. Back in 2006, another popular microcontroller, the Basic Stamp, cost nearly four times as much ($119) as an Arduino ($32). Today, an Arduino Uno costs just $22.

FIGURE 1-3: The original Arduino Serial board.

In one of my first Arduino workshops, I was told that the price was intended to be affordable for students. The price of a nice meal and a glass of wine at that time was about $42, so if you had a project deadline, you could choose to skip a nice meal that week and make your project instead.

The range of Arduino boards on the market is a lot bigger than it was back in 2006. In Chapter 2, you learn about just a few of the most useful Arduino and Arduino-compatible boards and how they differ to provide you with a variety of solutions for your own projects. Also, in Chapter 12, you learn all about a special type of circuit board called a *shield,* which can add useful, and in some cases phenomenal, features to your Arduino, turning it into a GPS (Global Positioning System) receiver, a mobile phone, or even a Geiger counter, to name just a few.

Learning by Doing

People have used technology in many ways to achieve their own goals without needing to delve into the details of electronics. Following are just a few related schools of thought that have allowed people to play with electronics.

Patching

Patching is a technique for experimenting with systems using wires. The earliest popular example of patching is in phone switchboards. For an operator to put you through to another line, he or she had to physically attach a cable.

This technique was also popular for synthesizing music, such as with the Moog synthesizer. When an electronic instrument generates a sound, it's really generating a voltage. Different collections of components in the instrument manipulate that voltage before it is outputted as an audible sound. The Moog synthesizer works by changing the path that that voltage takes, sending it through a number of different components to apply different effects.

Because so many combinations are possible, the musician proceeds largely through trial and error. But the simple interface means that this process is extremely quick and requires little preparation to get going.

Hacking

Hacking is a term that typically refers to the subversive use of technology. More generally, though, it refers to exploring systems and making full use of them or repurposing them to suit your needs.

Hacking in this sense is possible in hardware as well as software. A great example of hardware hacking is a keyboard hack. Say that you want to use a big red button to move through a slideshow. Most software programs contain keyboard shortcuts, and most PDF viewers move to the next page in a slideshow when the user presses the spacebar. If you know this, you ideally want a keyboard with only a spacebar.

Today's keyboards have a small circuit board, a bit smaller than a credit card (see Figure 1-4), containing lots of contacts that are connected when you press different keys. If you can find the correct combination, you can connect two contacts by using a pushbutton. Now every time you press that button, you send a space to your computer.

This technique is great for sidestepping the intricacies of hardware and getting the results you want. In the "Hacking Other Hardware" bonus chapter (`www.dummies.com/go/arduinofd`), you learn more about the joy of hacking and how you can weave hacked pieces of hardware into your Arduino project to control remote devices, cameras, and even computers with ease.

FIGURE 1-4: The insides of a keyboard, ready to be hacked.

Circuit bending

Circuit bending flies in the face of traditional education and is all about spontaneous experimentation. Children's toys are the staple diet of circuit benders, but really any electronic device has the potential to be experimented with.

By opening a toy or device and revealing the circuitry, you can alter the path of the current to affect its behavior. Although this technique is similar to patching, it's a lot more unpredictable. However, after you find a combination that produces a pleasing result, you can add or replace components, such as resistors or switches, to give the user more control over the instrument.

Most commonly, circuit bending is about sound, and the finished instrument becomes a rudimentary synthesizer or drum machine. Two of the most popular devices are the Speak & Spell (see Figure 1-5) and the Nintendo GameBoy. Musicians such as the Modified Toy Orchestra (`modifiedtoyorchestra.com`), in their own words, "explore the hidden potential and surplus value latent inside redundant technology." So think twice before putting your old toys on eBay!

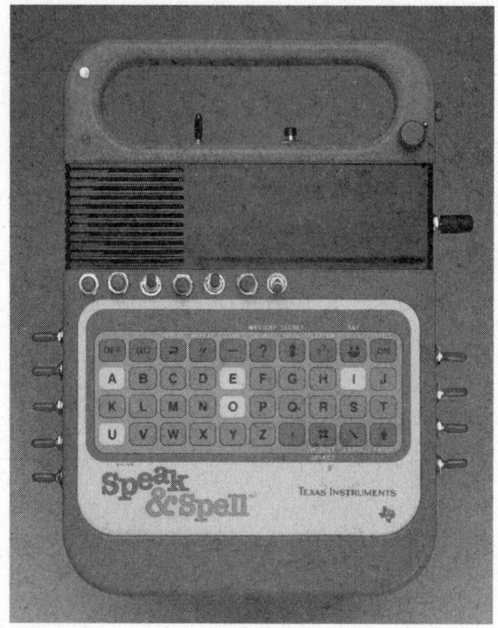

FIGURE 1-5: A Modified Toy Orchestra Speak & Spell after circuit bending.

Courtesy of Modified Toy Orchestra

Electronics

Although there are many ways to work around technology, eventually you'll want more of everything: more precision, more complexity, and more control.

If you learned about electronics at school, you were most likely taught how to build circuits using specific components. These circuits are based solely on the chemical properties of the components and need to be calculated in detail to make sure that the correct amount of current is going to the correct components.

These are the kind of circuits you find as kits at Radio Shack (or Maplin, in the United Kingdom) that do a specific job, such as an egg timer or a security buzzer that goes off when you open a cookie jar. These kits are good at their specific job, but they can't do much else.

This is where microcontrollers come in. When used with analog circuitry, microcontrollers can give that circuitry a more advanced behavior. They can also be reprogrammed to perform different functions as needed. Your Arduino is designed around one of these microcontrollers, and in Chapter 2, you look closely at an Arduino Uno to see exactly how it is designed and what it is capable of.

The microcontroller is the brains of a system, but it needs other electronic inputs and outputs to either sense or affect things in its environment.

Inputs

Inputs are senses for your Arduino. They tell it what is going on in the world. At its most basic, an input could be a switch, such as a light switch in your home. At the other end of the spectrum, it could be a gyroscope, telling the Arduino the exact direction it's facing in three dimensions. You learn all about basic inputs in Chapter 6, and more about the variety of sensors and when to use them in Chapter 11.

Outputs

Outputs allow your Arduino to affect the real world in some way. An output could be subtle and discreet, such as in the vibration of a cellphone, or it could be a huge visual display on the side of a building that can be seen for miles around. The first sketch in the book walks you through blinking an LED (see Chapter 3). From there, you can go on to controlling an electric motor (Chapter 7) and even controlling an LCD screen (Chapter 12).

Open Source

Open source software, in particular Processing, has had a huge influence on Arduino development. In the world of computer software, *open source* is a philosophy in which people share the details of a program and encourage others to use, remix, and redistribute it, as they like.

Just as the Processing software is open source, so are Arduino software and hardware. This means that the Arduino software and hardware are both released freely to be adapted as needed. You find the same spirit of openness also amongst the community on the Arduino forums.

On the official Arduino forums (`http://forum.arduino.cc/`) and many other ones around the world, people have shared their code, projects, and questions for an informal peer review. This sharing allows all sorts of people, including experienced engineers, talented developers, practiced designers, and innovative artists, to lend their expertise to novices in some or all of these areas. It also provides a means to gauge people's areas of interest, which occasionally filters into the official release of Arduino software or board design with refinements or additions. The Arduino website has an area known as the Playground (`playground.arduino.cc/`), where people are free to upload their code for the community to use, share, and edit.

This kind of philosophy has encouraged the relatively small community to pool knowledge on forums, blogs, and websites, thereby creating a vast resource for new Arduinists to tap into.

Despite the open-source nature of Arduino, a huge loyalty to Arduino as a brand exists — so much so that there is an Arduino naming convention of adding *-duino* or *-ino* to the name of boards and accessories (much to the disgust of Italian members of the Arduino team)!

» Discovering other Arduino boards

» Knowing where to shop for Arduinos

» Finding the right Arduino kit to get started

» Setting up a workspace

» Installing Arduino

» Introducing Arduino Web Editor

Chapter 2

Finding Your Board and Your Way Around It

The name *Arduino* encompasses a host of concepts. It can refer to an Arduino board, the physical hardware, the Arduino environment — that is, a piece of software that runs on your computer — and, finally, Arduino as a subject in its own right, as in this book: how the hardware and software can be combined with related craft and electronics knowledge to create a tool kit for any situation.

This chapter provides an overview of what you need to get started with Arduino. You may be eager to dive in, so you may want to quickly scan through this chapter, stopping at any areas of uncertainty and referring to them later as needed.

First, you learn about the components used on the Arduino Uno board, which is the starting point for most Arduinists. Beyond that, you get acquainted with the other available Arduino boards, how they differ, and what uses they have. The chapter lists major suppliers that can equip you with all the parts you need and examines some of the starter kits that are ideal for beginners and for accompanying this book. When you have the kit and a workspace, you're ready to start.

Lastly, you find out where to obtain the software to control your Arduino. I walk you through the steps for downloading and installing the software, and give you a brief tour of the environment in which you develop your Arduino programs.

Getting to Know the Arduino Uno

No one definitive Arduino board exists; many types of Arduino boards are available, each with its own design to suit various applications. Deciding what board to use can be a daunting prospect because the number of boards is increasing, each with new and exciting prospects. However, one board can be considered the archetype of the Arduino hardware, the one that almost all people start with and that is suitable for most applications. It's the Arduino Uno.

The most recent main board to date is the Arduino Uno R3 (released in 2011). Think of it as the plain vanilla of Arduino boards. It's a good and reliable workhorse suitable for a variety of projects. If you're just starting out, the Uno R3 is the board for you (see Figures 2-1 and 2-2).

FIGURE 2-1: The front of an Arduino Uno.

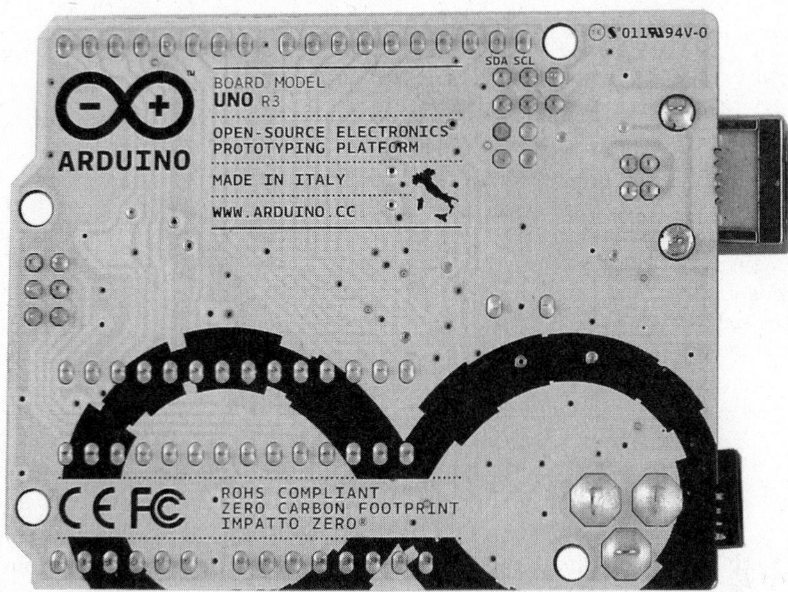

FIGURE 2-2: The back of an Arduino Uno.

Uno is Italian for the number one, named for the release of version 1.0 of the Arduino software. Predecessors to this release had a variety of names, such as Serial, NG, Diecimila (10,000 in Italian, to mark that 10,000 boards had been sold), and Duemilanove (2009 in Italian, the release date of the board); the Uno has ushered in some much needed order to the naming of the boards. R3 relates to the revision of the features on the board, which includes updates, refinements, and fixes. In this case, it is the third revision.

The board has many small components, described throughout much of this chapter.

The Brains: ATmega328P microcontroller chip

You can think of the microcontroller chip itself as the brains of the board. The chip used in the Arduino Uno is the ATmega328P, made by Atmel. It's the large, black component in the center of the board. This chip is known as an integrated circuit, or IC. It sits in a socket; if you were to remove it, it would look like the one shown in Figure 2-3.

FIGURE 2-3: An ATmega328P microcontroller all by itself.

This same chip can come in different forms, referred to as *packages.* The one in an Arduino Uno is in a *plated-through hole*, or PTH, package, named because of the way it makes contact with the board. Another variation you may find is the Arduino Uno SMD, or *surface mount device*, package, which is mounted on the surface of the board rather than in holes that go through it. This chip is much smaller, but it is not replaceable, whereas the PTH chip is. Apart from that, it functions exactly the same as the PTH and differs only in looks.

Header sockets

The *microcontroller socket* connects all the legs of the ATmega328 microcontroller chip to other sockets, referred to as *header sockets,* which are arranged around the edge of the board and are labeled for ease of use. These black sockets are divided into three main groups: digital pins, analog input pins, and power pins.

All these pins transfer a voltage, which can be either sent as output or received as an input. Why are these pins important? They allow you to connect additional circuitry to the board quickly and easily when prototyping with a breadboard (described in Chapter 6) and to design additional boards, called *shields,* that fit neatly on top of your Arduino board (see Chapter 12 for more on shields).

This same process of sending and receiving electrical signals is going on inside modern computers. But because they are so advanced and refined compared to a humble Arduino, it is difficult to directly link a computer accustomed to digital signals (0s and 1s) to an electronic circuit that deals with a range of voltages (0v to 5v in the ATmega328P's case).

The Arduino (see the sketch in Figure 2-4) is special because it can interpret these electric signals and convert them to digital signals that your computer can understand — and vice versa. It also enables you to write a program using software on a conventional computer that the Arduino IDE (integrated development environment) converts or compiles to electrical signals that your circuit can understand.

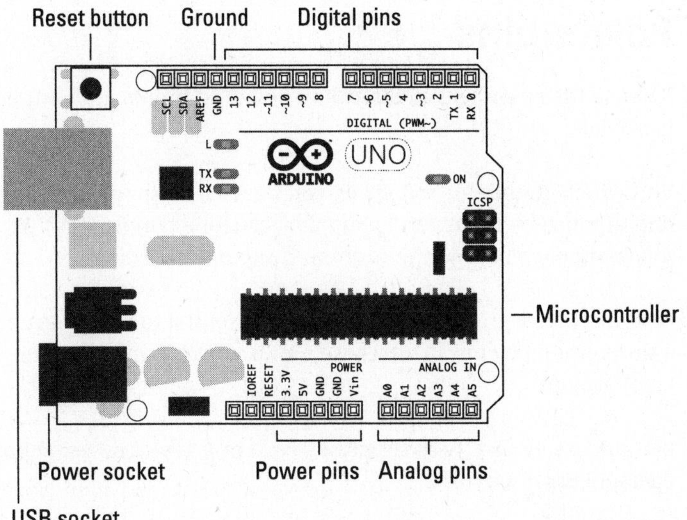

FIGURE 2-4: An Arduino Uno with all the important parts labeled.

By bridging this gap, it is possible to use a conventional computer's benefits — ease of use, user-friendly interfaces, and code that is easy for humans to understand — to control a wide range of electronic circuits and even give them complex behaviors with relative ease.

Digital pins

You use *digital pins* (refer to Figure 2-4) to send and receive digital signals. *Digital* implies that the pins have two states: off or on. In electrical terms, these states translate to a value of 0 or 5 volts, but no values in between.

Analog in pins

You use *analog in* pins (refer to Figure 2-4) to receive an analog value. An analog value is taken from a range of values. In this case, the range is the same 0V to 5V as with the digital pins, but the value can be at any point — 0.1, 0.2, 0.3, and so on.

What about analog out?

The shrewd ones among you may have noticed that there seem to be no *analog out* pins. In fact, the analog out pins are there, but they're hidden among the digital pins marked as PWM using the ~ symbol. PWM stands for *pulse-width modulation*, which is a technique you can use to give the impression of an analog output using digital pins. I explain how PWM works in Chapter 6. The ~ symbol appears next to digital pins 3, 5, 6, 9, 10, and 11, showing that you have six pins that are capable of PWM.

Power pins

You use the *power pins* to distribute power to inputs and outputs wherever power is needed.

Vin, which stands for *voltage in*, can be used to supply a voltage (V) equal to the one supplied by the external supply jack (for example, 12V). You can also use this pin to supply power to the Arduino from another source.

GND marks the ground pins, which are essential to complete circuits. There is also a third ground by pin 13. All these pins are linked and share the same (called *common*) ground.

You can use 5V or 3.3V to supply a 5-volt or 3.3-volt power supply, respectively, to components or circuits.

USB socket

To tell the microcontroller on the Arduino board what to do, you need to send a program to it. On the Uno, you send programs primarily by a USB connection. The large, metallic socket is a USB port for a USB A-B cable. This cable is similar to the one used on your home printer or scanner, so you may find a few around the house that can serve as handy spares. The Arduino uses the USB both for power and to transfer data. Using a USB cable is perfect for low-power applications and when data is being sent to or received from a computer.

Power socket

Next to the USB socket is another socket; this one is for power. This socket allows you to power your Arduino from an external power supply. The supply could be from an AC-to-DC adaptor (similar to those used on other consumer electronics), a battery, or even a solar panel.

The connector needed is a 2.1 mm center positive plug. *Center positive* simply means that the plug has an outside and an inside that fit the socket and that the inside of the plug must be positive. You should be able to find this plug among the standard connectors that come with most power supplies; otherwise, you can buy the connector and attach it to bare wires.

WARNING

If you connect a power supply that is the opposite (center negative), it is known as having a reverse polarity. Components on the Arduino Uno will resist your attempts to send voltage the wrong way around the board, but those components can melt in the process of saving your board, depending on how much power you're sending and how long it takes you to notice the burning smell! If you reverse the polarity when using the Vin, 5V, or 3.3V pins, you bypass this protection and almost instantly destroy several parts of your board and the ATmega328P chip.

The recommended voltage for the Uno board is 7–12V. If you supply too little power, your board might not function correctly. Or if you provide too much power, your board might overheat and potentially be damaged.

LEDs

LEDs come in a variety of shapes and sizes and are found in almost every modern piece of consumer electronics, from bike lights to TVs to washing machines. You see a lot more of them in numerous examples throughout the book.

The components described in this section are tiny. The Uno board has four LEDs (light-emitting diodes) labeled L, RX, TX, and ON. An *LED* is a component that produces light when electrical current flows through it.

These four LEDs are used to indicate activity on the board, as follows:

- ON is green and signifies that your Arduino is powered.
- RX and TX tell you that data is being **r**eceived or **t**ransmitted by the board.
- L is a special LED connected to digital pin 13. It is great for testing to see whether your board is functioning as you want.

If your Arduino is plugged in but you don't see any lights, double-check that:

- Your USB cable is plugged in
- Your USB port is working — try another device in the port
- Your cable is working — try another cable, if possible

If the LED still doesn't illuminate, something is probably wrong with your Arduino. Your first destination should be the Arduino troubleshooting page at `http://arduino.cc/en/Guide/troubleshooting`. If you still have no luck, request a replacement Arduino from where you purchased the device.

Reset button

The Uno board also has a button next to the USB socket called the reset button. It resets the program on the Arduino or stops it completely when held down for a time. Connecting a wire between GND and the reset pin, which is located next to the 3.3V, achieves the same results.

The board has many other components, all of which perform important jobs, but the ones described in this section are the key ones for you to know for now.

Discovering Other Arduino Boards

The preceding section describes the standard USB Arduino board, but you should be aware that many others exist, all designed with different needs in mind. Some offer more functionality, and others are designed to be more minimal, but generally they follow a design similar to that of the Arduino Uno. For this reason, all examples in this book are based on the Uno.

Previous revisions of the Uno should work without any changes, but if you're using an older or more specialized board, be sure to follow instructions specific to it. This section gives you a brief rundown of other available boards.

Official Arduino boards

Although Arduino is open source, it is also a trademarked brand, so to guarantee the quality and consistency of its products, the Arduino team must properly approve new boards before they are officially recognized and can bear the name Arduino. You can recognize official boards first by the name — Arduino Pro, Arduino Fio, or Arduino LilyPad, for example. Other nonofficial boards often include *Arduino compatible* or *for Arduino* in the name. The other way to recognize an official Arduino, made by the Arduino team, is by the branding (in the most recent versions): They are turquoise and display the infinity symbol somewhere on the board, along with a link to `www.arduino.cc`. Some other companies also have their boards accepted as official boards, so you may find other company names printed on them, such as Adafruit Industries and SparkFun.

Because the schematics for the Arduino board are open source, unofficial Arduino boards have a lot of variation, which people have made for their own needs. These boards are usually based on the same microcontroller chips to remain compatible with the Arduino software, but they require extra consideration and reading to be sure that they will work as expected. The Seeeduino v4.2 (by Seeed Studio), for example, is based on the Arduino Uno and is 100 percent compatible but adds various extra connections, switches, and sockets, which may be of more use to you than an official Arduino board in certain situations.

Official boards are the safe option for beginners to choose because the majority of Arduino examples online are based on these boards. Because of this, official boards are more widely used, and because of *that*, any errors or bugs in the board design are likely to be remedied with the next revision or at least well documented.

Arduino Leonardo

The Leonardo has the same *footprint* (shape of circuit board), but the microcontroller used is different and can be recognized as a keyboard or mouse by a computer. In the "Hacking Other Hardware" bonus chapter (`www.dummies.com/go/arduinofd`), I provide more detail about the differences between this board and the Uno and how to use it.

Arduino Mega 2560 R3

As the name suggests, the Arduino Mega 2560 R3 is a bigger board than the Arduino Uno R3. It is for people who want more: more inputs, more outputs, and more processing power! The Mega has 54 digital pins and 16 analog pins whereas the Uno has a measly 15 digital and 6 analog pins.

Arduino Mega ADK

The Arduino Mega ADK is essentially the same board as the Mega 2560 Arduino but is designed to interface with Android phones. This means you can share data between your Android mobile or tablet and an Arduino to broaden the range of either.

Arduino Nano 3.0

The Arduino Nano 3.0 is a condensed Arduino that measures just 0.70" x 1.70". This size is perfect for making your project smaller. The Nano has all the power of an Arduino Uno (both use the same ATmega328 microcontroller), but it's a fraction of the size. The Nano also handily fits into a breadboard, making it ideal for prototyping as well.

Arduino Mini R5

Despite what the names suggest, the Arduino Mini R5 is smaller than the Nano. This board also uses the same ATmega328 microcontroller chip but is condensed further, removing all header pins and the Mini-USB connector of the Nano. The Mini board is great if space is at a premium, but it does require great care when connecting because an incorrect connection can easily destroy the board.

Arduino 101

The Arduino 101 uses the Intel Curie module to provide a board that's as easy to use as Arduino but with the latest Bluetooth low energy technology. It also features a six-axis accelerometer to detect movement, making it a great companion for any wearable technology projects or for talking to other Bluetooth devices.

Arduino MKR1000

The Arduino MKR1000 is a simple but powerful board, combining a 32-bit processor with a Wi-Fi module. It's perfect for linking sensors and actuators to the Internet or prototyping your next Internet of Things (IoT) project. Along with its small form factor, it has a handy battery connector to allow you to take your project on the move.

Contributed (approved) Arduinos

Many Arduino boards are now standardized and designed by the Arduino team. However, some have been contributed by other companies, such as Adafruit Industries and SparkFun, and are recognized as official boards. I list a few of the best ones here.

LilyPad Arduino USB

The LilyPad Arduino USB was made for projects in which technology is combined with textiles to aid in the development of e-textiles or wearable electronics projects. The LilyPad and its accompanying *breakout boards* (printed circuit board that make it easy to integrate various components without the need to build your own boards) can be sewn together using conductive thread instead of conventional wire. This board was designed and developed by Leah Buechley of MIT (`www.leahbuechley.com/`) and SparkFun Electronics. If you're interested in e-textiles or wearable electronics, check out the excellent tutorial on SparkFun's site that introduces the latest version of the board and the ProtoSnap kit. (Go to `learn.sparkfun.com/tutorials/lilypad-development-board-hookup-guide`.)

Arduino Pro

The Arduino Pro is a minimal and super skinny Arduino, by SparkFun Electronics, based on the same microcontroller as those used in the Uno. It comes without any of the normal headers or sockets but has all the same capabilities of an Uno. It's ideal when height is at a short supply. In addition, it has a battery socket, which allows you to easily make your project portable.

Arduino Pro Mini

The Pro Mini is another SparkFun product; this one pushes the minimalism of the Arduino Pro to new limits. In the scale of Arduinos, the Pro Mini sits neatly between the Nano and the Mini. It has none of the header pins or the Mini-USB port of the Nano, and is slightly more spread out than the Arduino Mini. The Pro Mini has none of the safety features of the Uno, so take great care when wiring because a wrong connection can easily destroy the board.

Shopping for Arduino

Initially, Arduino boards were available only from a small number of hobby shops scattered across the world. Now you have lots of places to purchase an Arduino, as listed in the following sections. Later in this chapter, I tell you about beginner's kits, which pull together the basic components for you; I recommend a kit to get you started on your Arduino endeavors.

On the Arduino site, you can find an exhaustive list of Arduino resellers and manufacturers around the world (`store.arduino.cc/distributors`). Here are just a few of the most useful.

Arduino Store

`https://store.arduino.cc/usa/`

A good place to start is the Arduino Store. This location has all the latest Arduino boards, kits, and a select few components.

Adafruit

`www.adafruit.com/`

MIT engineer Limor "Ladyada" Fried founded Adafruit in 2005. Through its website, the company offers a wealth of resources, including products that the company designs and makes itself; other products sourced from all over; tools and equipment; and tutorials, forums, and videos covering a wide range of topics. Adafruit is based in New York, New York (it's a wonderful town!). It distributes worldwide.

SparkFun

www.sparkfun.com/

SparkFun sells all sorts of parts for every variety of electronics projects. As well as selling Arduino-compatible hardware, it designs and makes a lot of its own boards and kits. SparkFun has an excellent site that acts as a shop front, a support desk, and a classroom for Arduinoists. SparkFun also has active (and vocal) commenters on each of its product pages, which help to support and continually improve the company's products. SparkFun was founded in 2003 and is based in Boulder, Colorado.

Seeed Studio

www.seeedstudio.com

Based in Shenzhen, China, Seeed Studio is self-described as an "open hardware facilitation company." The shop uses local manufacturing to quickly make prototypes and small-scale projects that are distributed worldwide. In addition to manufacturing and selling products, the company offers a community area on its website where people can vote for the projects that they want Seeed Studio to bring to fruition (www.seeedstudio.com/wish/).

Watterott Electronic

www.watterott.com

Founded by Stephan Watterott back in 2008, Watterott Electronic, based in Leinefelde, Germany, began as a one-man business providing a few microcontroller boards and specialist components. A lot has changed since then, and the company now stocks a large array of Arduino-compatible boards, components and other hardware.

Electronics distributors

Many well-established global electronics distribution companies deliver to all corners of the world. Relatively recently, they have started stocking Arduino boards, but they are especially useful for bulk buying of components when you know what you're looking for.

TIP

You can lose days searching through the extensive catalogues of components, so it's always a good idea to know the name of what you're looking for before you start!

Here are some global distribution companies that stock Arduino boards and components:

- **Digi-Key:** www.digikey.com
- **Farnell:** www.farnell.com
- **Mouser:** www.mouser.com
- **Rapid:** www.rapidonline.com
- **RS Components:** www.rs-components.com

Amazon

So much has the popularity of Arduino grown that it's also now stocked at Amazon (www.amazon.com). Most Arduino boards as well as a variety of components and kits are available here, although they're harder to find than on more hobby-specific sites.

eBay

Originally a source of only second-hand components, eBay (www.ebay.com) has now become a platform used by many retailers and wholesalers. Here you will find official boards, custom boards, and popular components. Some imitation Arduino boards are also available. Most are harmless and will work the same as an official board, but the quality control is often worse, which can result in strange behavior or damage to a connected devices. If in doubt, use a genuine Arduino board from a certified distributor.

Kitted Out: Starting with a Beginner's Kit

By this point, you probably know a bit about the Arduino board, but no board is an island; you need lots of other bits so that you can make use of the board. In the same way that a computer is of no use without a mouse and keyboard, an Arduino is of no use — or at least not as much fun — without components.

Every new Arduinist should follow several basic examples to learn the fundamentals of Arduino (which this book covers in Chapters 3 through 7). These examples are all achievable with a few basic components. To save you the time and effort of finding these components, a few enterprising individuals and companies have put together kits that let you experiment in no time!

Many kits have been designed by different individuals and companies based on their experiences, likes, and dislikes. You can also find a lot of components that do the same job but have different appearances, based on their application.

The following short list describes a few core components that should be included in all good Arduino beginners' kits:

- **Arduino Uno:** This is the board you know and love.
- **USB A-B cable:** This cable is essential to make use of your Arduino. It can also be found on printers and scanners.
- **LEDs:** Light-emitting diodes in various colors are great for providing visual feedback for your project as well as for testing lighting projects on a small scale.
- **Resistors:** Also referred to as *fixed resistors,* these are fundamental electrical components used to resist the flow of current through a circuit. Resistors are essential for the smooth running of most circuits. Each resistor has a fixed value, which is indicated by a colored band on the side of the resistor. These bands help you to quickly identify a resistor's resistance visually.
- **Variable resistors:** Also known as *potentiometers* or *pots,* variable resistors resist current in the same way as fixed-value resistors, but they can change their resistance. They are most commonly used in radios and hi-fi equipment for tuning and volume control dials, and are available also in other housings to detect other inputs such as force or flex on a surface.
- **Diodes:** Also known as *rectifier diodes,* diodes are similar to LEDs but without the light. They have an extremely high resistance to the flow of current in one direction and an extremely low (ideally zero) resistance in the other, which is the same reason that an LED works in only one direction. Instead of emitting light like LEDs, diodes control the flow of current throughout your circuit.
- **Photo diodes:** Also known as *photo resistors* or *light-dependent resistors* (LDRs), photo diodes change their resistance when light falls on them. They can have a variety of different uses depending on how they're placed relative to the light source.
- **Pushbuttons:** These components are found behind the scenes in many bits of consumer electronics such as game console controllers and stereos. They're used to either connect or disconnect parts of a circuit so that your Arduino can monitor human inputs.
- **Temperature sensors:** These sensors tell you what the ambient temperature is wherever they are placed. They are great for observing changes in your environment.

- **Piezo buzzer:** A piezo buzzer is technically described as a discrete sounding device. These simple components can be supplied with a voltage to produce simple notes or music. They can also be attached to surfaces to measure vibrations.
- **Relays:** These electrically operated switches are used to switch higher power circuits using your low-voltage Arduino. Half of a relay is an electromagnet, and the other half is a magnetic switch. The electromagnet can be activated by the 5V of the Arduino, which moves the contact of the switch. Relays are essential for bigger lighting and motor-based projects.
- **Transistors:** These components are the basis for all modern computers. Transistors are electrically operated switches, similar to relays, but the switch happens on a chemical level rather than a physical level. This means that the switching can be super fast, making transistors perfect for high-frequency operations such as animating LED lighting or controlling the speed of motors.
- **DC motors:** These motors are simple electric motors. When electric current is passed through a motor, it spins in one direction; when that direction is reversed, it spins in the other. Electric motors come in great variety, from those in your phone that vibrate to those in electric drills.
- **Servo motors:** These motors have on-board circuitry that monitors their rotation. Servo motors are commonly used for precision operations such as the controlled opening of valves or moving the joints of robots.

Here are a few of the better-known kits, ascending in price. They include all the components in the preceding list, and any will be an excellent companion for the examples in this book:

- Arduino Starter Kit from Arduino, available from `store.arduino.cc/usa/arduino-starter-kit`, costing $87.90 at the time of this writing
- Starter Kit for Arduino (ARDX) by Oomlout, available from `www.adafruit.com/products/170`, costing $84.95 at the time of this writing
- SparkFun Inventor's Kit 4.0 by SparkFun, available from `www.sparkfun.com/products/14265`, costing $99.95 at the time of this writing

You can create all the basic examples in this book with any kit in the preceding list, although a slight variation may occur in the number and type of components. See an example Arduino kit in Figure 2-5. Sometimes the same component can take many different forms, so be sure to carefully read the parts list to make sure that you can identify each of the components before you start. Cheaper kits are available, but these will likely not include some components, such as motors or the variety of sensors.

FIGURE 2-5: An example of an Arduino kit with an Arduino Uno and a good range of components.

Preparing a Workspace

When working on an Arduino project, you could sit on your sofa or be at the top of a ladder. I've been there before! But just because it's possible to work this way doesn't mean that it's sensible or advisable. You're far better off to have a good workspace before you dive into your experiment, especially when you're just starting out with Arduino.

Working with electronics is a fiddly business. You're dealing with lots of tiny, delicate, and very sensitive components, so you need great precision and patience when assembling your circuit. If you're in a dimly lit room trying to balance things on your lap, you'll quickly go through your supply of components by either losing or destroying them.

It's always good to make life easy for yourself, and the best way to do this when working on Arduino projects is to prepare your workspace. The ideal workspace has the following:

- Large, uncluttered desk or table
- Good work lamp

- Comfortable chair
- Cup of tea or coffee (recommended)

After you're comfortable, you're ready to learn how to set up the Arduino software. The Arduino software is a type of an integrated development environment (IDE), a tool common in software development that allows you to write, test, and upload programs. Versions of Arduino software are available for Windows, macOS, Linux, and Arduino Web Editor.

Installing Arduino

This section talks you through installing the Arduino environment on your platform of choice. These instructions are specifically for installation using an Arduino Uno, but also work just as well for previous boards.

Arduino is free to download from `www.arduino.cc/en/main/software` and is supported on macOS, Windows 32-bit and 64-bit, and Linux 32-bit, 64-bit, and ARM. At the time of this writing, Arduino was version 1.8.4.

To install Arduino, do the following:

- **On a Mac:** The .zip file unzips automatically, producing the Arduino app file, which you can then drag to your Applications folder. From there, you can drag Arduino to the dock for easy access or create a desktop alias.
- **On Windows:** Download the installer file and double-click to install the file in a directory of your choice. You also have the option to create shortcuts and install the latest USB drivers.
- **On Linux:** Download the compressed file for your version of Linux and extract the package. Open the Arduino folder (whose name includes the version number, such as arduino-1.8.4) created by the extraction process and find the install.sh file. Right-click it and choose Run in Terminal from the contextual menu.

Surveying the Arduino Environment

Programs written for Arduino are known as *sketches*. This naming convention was passed down from Processing, which allowed users to create programs quickly, in the same way that you would scribble an idea in a sketchbook.

Before you look at your first sketch, I encourage you to explore the Arduino software. The Arduino software is an integrated development environment, or IDE, and this environment is presented to you as a graphical user interface, or GUI (pronounced "goo-ey").

A *GUI* provides a visual way of interacting with a computer. Without it, you would need to read and write lines of text, similar to what you may have seen in the DOS prompt in Windows, Terminal in macOS, and in that bit about the white rabbit at the start of *The Matrix.*

The turquoise window is Arduino's GUI. It's divided into the following four main areas (labeled in Figure 2-6):

- **Menu bar:** Similar to the menu bar in other programs you're familiar with, the Arduino menu bar contains drop-down menus to all the tools, settings, and information that are relevant to the program. In macOS, the menu bar is at the top of your screen; in Windows and Linux, the menu bar is at the top of the active Arduino window.

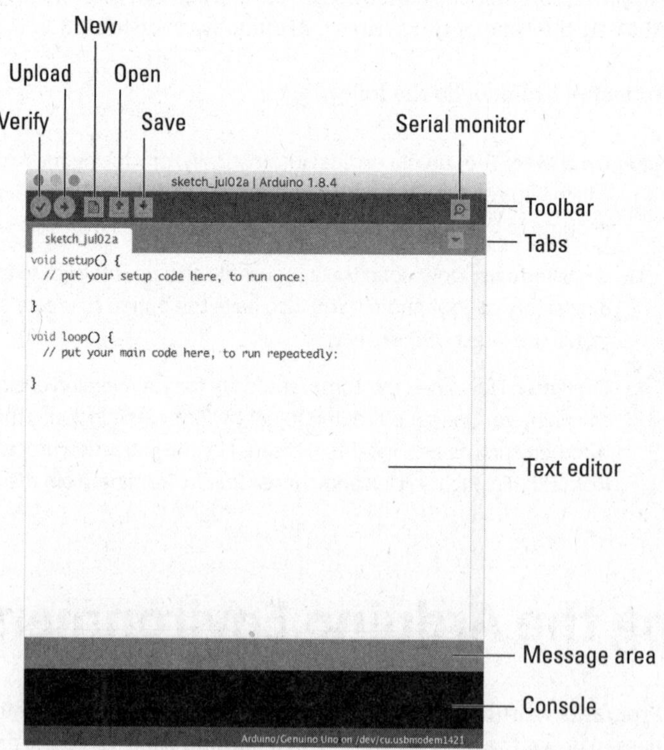

FIGURE 2-6: The areas of the GUI.

- **Toolbar:** The toolbar contains several buttons that are commonly needed when writing sketches for Arduino. These buttons, which are also available on the menu bar, perform the following functions:
  - *Verify:* Checks that your code makes sense to the Arduino software. Known as *compiling*, this process is a bit like a spelling and grammar checker. Be aware, however, that although the compiler checks that your code has no obvious mistakes, it does not guarantee that your sketch works correctly.
  - *Upload:* Sends your sketch to a connected Arduino board. It automatically compiles your sketch before uploading it.
  - *New:* Creates a sketch.
  - *Open:* Opens an existing sketch.
  - *Save:* Saves the current sketch.
  - *Serial monitor:* Allows you to view data that is being sent to or received by your Arduino board.
- **Text editor:** This area displays your sketch as text. It is almost identical to a regular text editor but has a few added features. Some text is color coded if the Arduino software recognizes it. You also have the option to auto-format the text so that it is easier to read.
- **Message area:** Even after years of using Arduino, you'll still make mistakes (everybody does), and the message area is the first way for you to find out that something is wrong. (***Note:*** The second way is the smell of burning plastic.)

Using Arduino Web Editor

Arduino Web Editor is a cloud-based version of the downloadable IDE and a great way to get started quickly.

Cloud-based means that Web Editor runs in your browser and saves your project files to your online account instead of on your computer. Apart from those differences, Web Editor has all the core features of the downloadable IDE but is more flexible, working across platforms without the need to install any files. In Figure 2-7, you can see how the layout differs.

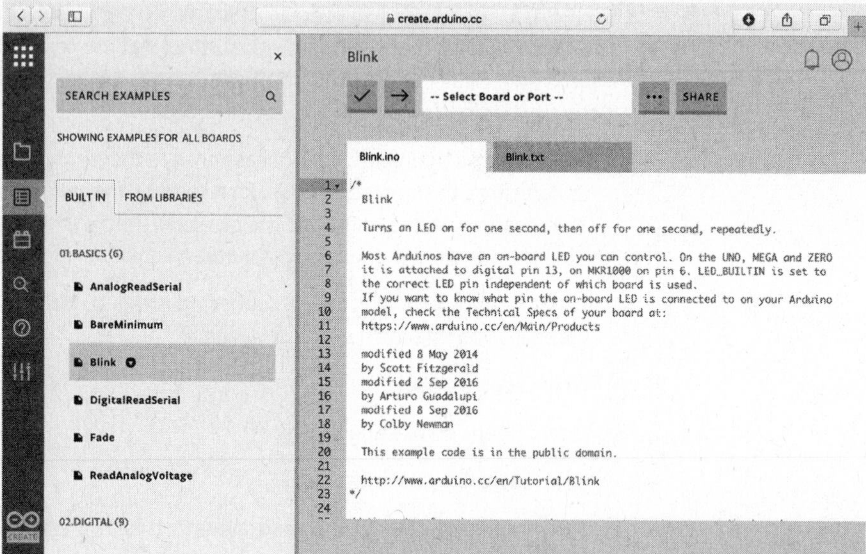

FIGURE 2-7: The areas of the Web Editor GUI.

» Setting the software

» Uploading Blink

» Completing your first Arduino sketch

Chapter 3

Blinking an LED

Brace yourself. You are about to take your first real step into the world of Arduino! You've bought a board, maybe an Arduino starter kit (possibly from one of the suppliers I recommended), and you're ready to go.

It's always a good idea to have a clear work surface or desk to use when you're tinkering. It's not uncommon to drop or misplace some of the many tiny components you work with, so make sure your workspace is clear, well lit, and accompanied by a comfortable chair.

By its nature, Arduino is a device intended for performing practical tasks. The best way to learn about Arduino, then, is in practice — by working with the device and *doing* something. That is exactly the way I write about it throughout this book. In this chapter, I take you through some simple steps to get you on your way to making something.

I also walk you through uploading your first Arduino sketch. After that, you examine how it works and see how to change it to do your bidding.

Working with Your First Arduino Sketch

In front of you now should be an Arduino Uno, a USB cable, and a computer running your choice of operating system (Windows, macOS, Linux, or Web Editor). The next section shows what you can do with this little device.

To make sure that the Arduino software is talking to the hardware, you upload a sketch. What is a sketch, you ask? Arduino was created as a device that allows people to quickly prototype and test ideas using little bits of code that demonstrate the idea — kind of like how you might sketch out an idea on paper. For this reason, programs written for Arduino are referred to as *sketches*. Although a device for quick prototyping was its starting point, Arduino devices are being used for increasingly complex operations. So don't infer from the name sketch that an Arduino program is trivial.

The specific sketch you want to use here is called Blink. It's about the most basic sketch you can write, a sort of "Hello, world!" for Arduino. Click in the Arduino window. From the menu bar, choose File ⇨ Examples ⇨ 01.Basics ⇨ Blink (see Figure 3-1).

FIGURE 3-1: Find your way to the Blink sketch.

A new window opens in front of your blank sketch and looks similar to Figure 3-2.

Identifying your board

Before you can upload the sketch, you need to check a few things. First you should confirm which board you have. As I mention in Chapter 2, you can choose from a variety of Arduino devices and several variations on the USB board. The latest generation of USB boards is the Uno R3. If you bought your device new, you can be fairly certain that you have this type of board. To make doubly sure, check the back of the board. You should see details about the board's model, as shown in Figure 3-3.

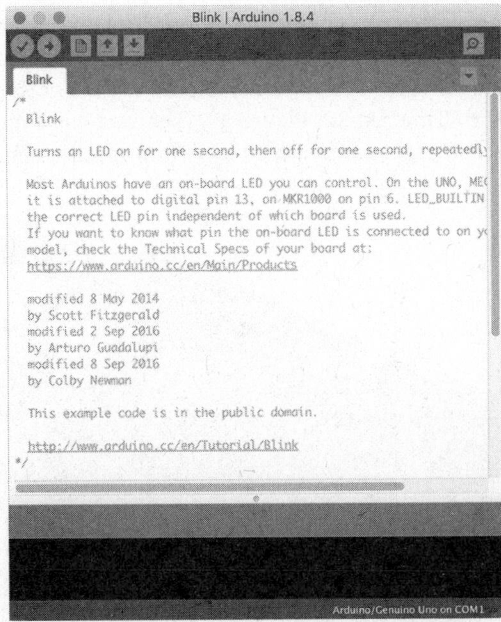

FIGURE 3-2: The Arduino Blink sketch.

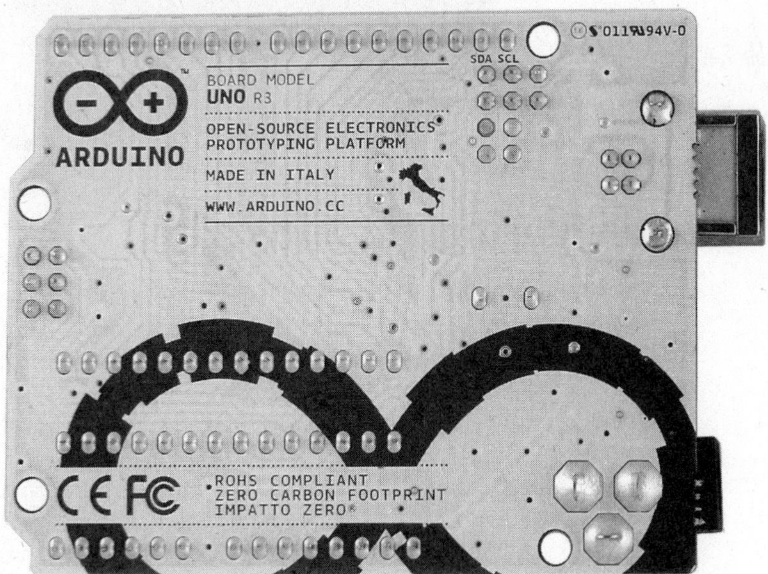

FIGURE 3-3: The back side of the Arduino Uno.

Also worth checking is the Atmel chip on the Arduino. As I mention in Chapter 2, the Atmel chip is the brains of the Arduino and is similar to the processor in your computer. Because the Uno and earlier boards allow you to replace the chip, there is always a chance, especially with a used board, that the chip has been replaced with a different one.

Although the Atmel chip looks distinctive on an individual board, if you compare it to an older Arduino, telling them apart at first glance would be difficult. The important distinguishing feature is written on the surface of the chip. In this case, you are looking for ATmega328P. Figure 3-4 shows a close-up of the chip. The chip has *PU* at the end of its name, which is a more specific name for this variant.

FIGURE 3-4: Close-up of the ATmega328P-PU chip.

Configuring the software

After you confirm the type of board you're using, you have to provide that information to the software. From the Arduino main menu bar (at the top of the Arduino window in Windows and at the top of the screen in macOS), choose Tools ⇨ Board. You should see a list of the different kinds of boards supported by the Arduino software. Select your board from the list, as shown in Figure 3-5.

Next, you need to select the serial port. The *serial port* is the connection that enables your computer and the Arduino device to communicate. *Serial* describes the way that data is sent, one bit of data (0 or 1) at a time. The *port* is the physical interface, in this case a USB socket. I talk more about serial communication in Chapter 6.

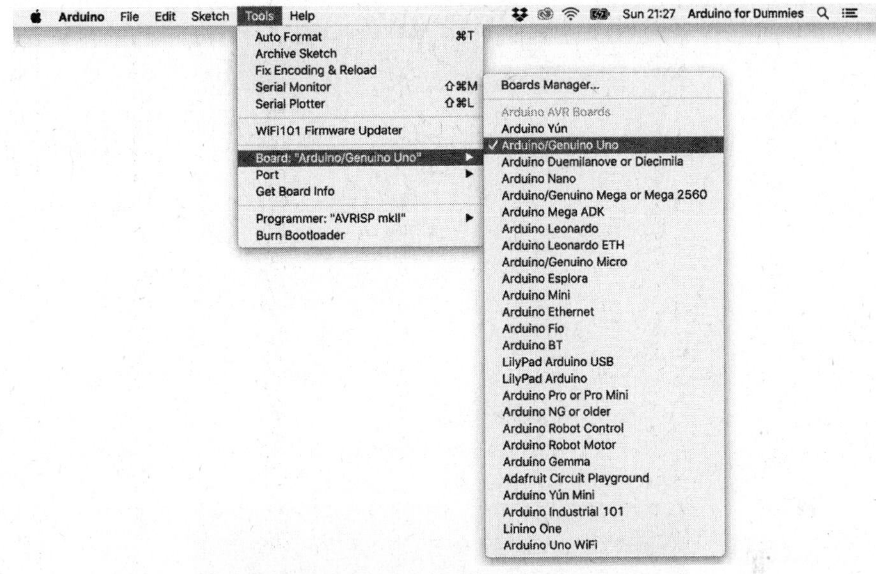

FIGURE 3-5: Select Arduino/Genuino Uno from the Board menu.

To determine the serial port, choose Tools ➪ Port. A list of devices connected to your computer is displayed (see Figure 3-6). This list contains any device that can talk in serial, but for the moment, you're interested only in finding the Arduino. If you've just installed Arduino and plugged it in, it should be at the top of the list. For macOS users, the Arduino is shown as /dev/cu.usbmodem*XXXXXX* or /dev/cu.usbmodem*XXXXXX* (where *XXXXXX* is a randomly assigned number). On Windows, the same is true, but the serial ports are named COM1, COM2, COM3, and so on. The highest number is usually the most recent device.

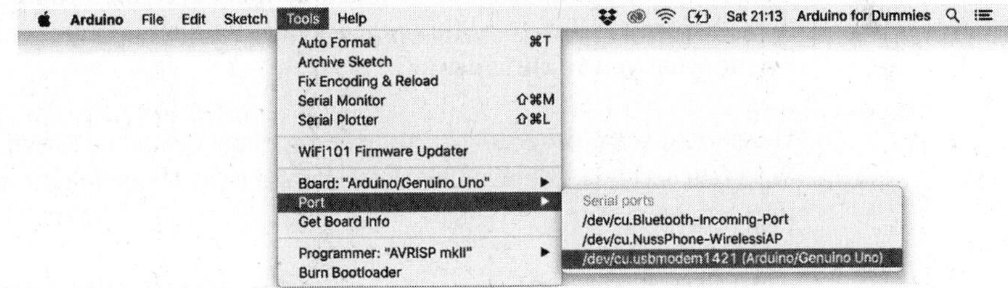

FIGURE 3-6: A list of serial connections available to the Arduino software.

After you find your serial port, select it. It should appear in the bottom right of the Arduino graphic user interface (GUI), along with the board you selected (see Figure 3-7).

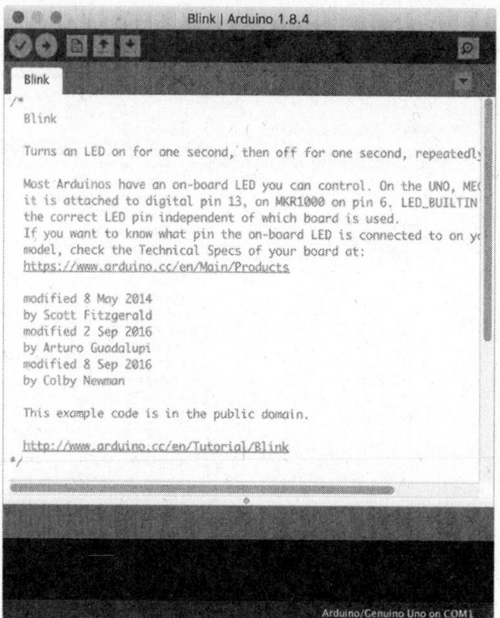

FIGURE 3-7: The Arduino GUI board and port.

Uploading the sketch

Now that you have told the Arduino software what kind of board you're communicating with and which serial port connection it's using, you can upload the Blink sketch you found earlier in this chapter.

First click the Verify button (check mark). Verify checks the code to make sure it makes sense, that is, that the syntax is written in a way Arduino can understand (see Chapter 2). However, this process doesn't necessarily mean that your code will do what you're anticipating.

You should see a progress bar and the message *Compiling Sketch* (see Figure 3-8) for a few seconds, followed by the message *Done Compiling* after the process has finished.

If the sketch compiled successfully, you can now click the Upload button (right arrow) next to the Verify button. A progress bar appears, and you see a flurry of activity on your board from the two LEDs marked RX and TX (mentioned in Chapter 2). These show that the Arduino is sending and receiving data. After a few seconds, the RX and TX LEDs stop blinking, and a *Done Uploading* message appears at the bottom of the Arduino window (see Figure 3-9).

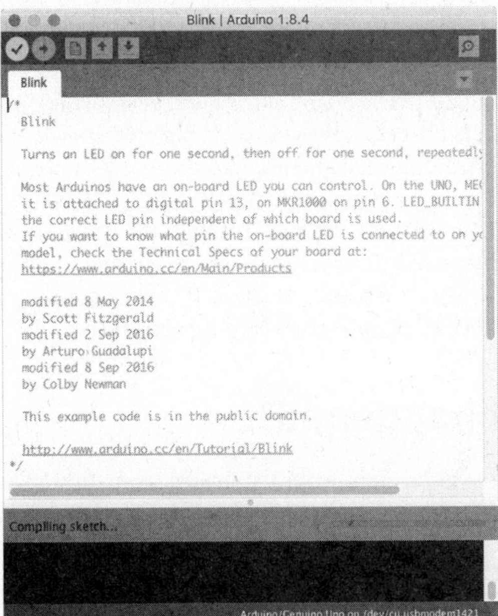

FIGURE 3-8: The progress bar shows that the sketch is compiling.

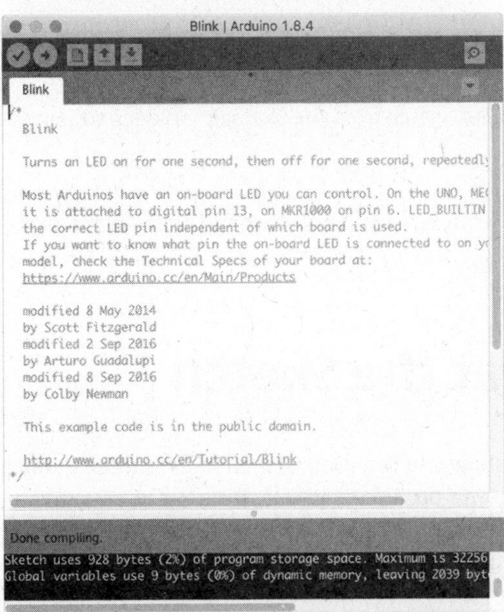

FIGURE 3-9: The Arduino software has successfully uploaded a sketch.

Congratulate yourself!

You should see the LED marked L blinking away reassuringly: on for a second, off for a second. If that is the case, give yourself a pat on the back. You've just uploaded your first piece of Arduino code and entered the world of physical computing!

If you don't see the blinking L, go back through the previous sections. Make sure you have installed Arduino properly and then give it one more go. If you still don't see the blinking L, check out the excellent troubleshooting page on the official Arduino site at `arduino.cc/en/guide/troubleshooting`.

What just happened?

Without breaking a sweat, you've just uploaded your first sketch to an Arduino. Congratulations!

Just to recap, you have now:

- Plugged your Arduino into your computer
- Opened the Arduino software
- Set the board and serial port
- Opened the Blink sketch from the Examples folder and uploaded it to the board

In the following section, I walk you through the various parts of the sketch you just uploaded.

Looking Closer at the Sketch

In this section, I show you the Blink sketch in a bit more detail so that you can see what's actually going on. When the Arduino software reads a sketch, it quickly works through it one line at a time, in order. So the best way to understand the code is to work through it the same way but slowly.

Arduino uses the programming language C, which is one of the most widely used languages of all time. It's an extremely powerful and versatile language, but it takes some getting used to.

If you followed the previous section, you should already have the Blink sketch on your screen. If not, you can find it by choosing File ➪ Examples ➪ 01.Basics ➪ Blink (refer to Figure 3-1).

When the sketch is open, you should see something like this:

```
/*
  Blink
  Turns on an LED on for one second, then off for one second, repeatedly.

  Most Arduinos have an on-board LED you can control. On the UNO, MEGA and ZERO
  it is attached to digital pin 13, on MKR1000 on pin 6. LED_BUILTIN is set to
  the correct LED pin independent of which board is used.
  If you want to know what pin the on-board LED is connected to on your Arduino
   model, check
  the Technical Specs of your board at https://www.arduino.cc/en/Main/Products

  This example code is in the public domain.

  modified 8 May 2014
  by Scott Fitzgerald

  modified 2 Sep 2016
  by Arturo Guadalupi

  modified 8 Sep 2016
  by Colby Newman
*/

// the setup function runs once when you press reset or power the board
void setup() {
  // initialize digital pin LED_BUILTIN as an output.
  pinMode(LED_BUILTIN, OUTPUT);
}

// the loop function runs over and over again forever
void loop() {
  digitalWrite(LED_BUILTIN, HIGH);   // turn the LED on (HIGH is the voltage level)
  delay(1000);                       // wait for a second
  digitalWrite(LED_BUILTIN, LOW);    // turn the LED off by making the voltage LOW
  delay(1000);                       // wait for a second
}
```

The sketch is made up of lines of code. When looking at the code as a whole, you can identify three distinct sections:

- Comments
- void loop
- void setup

Read on for more details about each of these sections.

Comments

Here's what you see in the first section of the code:

```
/*
  Blink
  Turns on an LED on for one second, then off for one second, repeatedly.

  Most Arduinos have an on-board LED you can control. On the UNO, MEGA and ZERO
  it is attached to digital pin 13, on MKR1000 on pin 6. LED_BUILTIN is set to
  the correct LED pin independent of which board is used.
  If you want to know what pin the on-board LED is connected to on your Arduino
   model, check
  the Technical Specs of your board at https://www.arduino.cc/en/Main/Products

  This example code is in the public domain.

  modified 8 May 2014
  by Scott Fitzgerald

  modified 2 Sep 2016
  by Arturo Guadalupi

  modified 8 Sep 2016
  by Colby Newman
*/
```

Multiline comment

Note that the code lines are enclosed within the symbols `/*` and `*/`. These symbols mark the beginning and end of a *multiline* or *block comment*. Comments are written in plain English and, as the name suggests, provide an explanation or comment on the code. The software ignores comments when the sketch is compiled and uploaded. Consequently, comments can contain useful information about the code without interfering with how the code runs.

In this example, the comment simply tells you the name of the sketch and what it does, and provides a note explaining that this example code is in the public domain. Comments often include other details such as the name of the author or editor, the date the code was written or edited, a short description of what the code does, a project URL, and sometimes even contact information for the author.

Single-line comment

Further down the sketch, inside the `setup` and `loop` functions, text on your screen appears with the same shade of gray as the comments above. This text is also a comment. The symbols `//` signify a single-line comment as opposed to a multi-line comment. Any code written after these forward slashes will be ignored for that line. In this case, the comment is describing the piece of code that comes after it:

```
// the setup function runs once when you press reset or power the board
```

Functions

The next two sections are functions and begin with the word `void`: `void setup` and `void loop`. A *function* is a bit of code that performs a specific task, and that task is often repetitive. Rather than writing the same code out again and again, you can use a function to tell the code to perform the task again.

Consider the general process you follow to assemble IKEA furniture. If you were to write these general instructions in code, using a function, they would look something like this:

```
void buildFlatpackFurniture() {
          buy a flatpack;
          open the box;
          read the instructions;
          put the pieces together;
          admire your handiwork;
          vow never to do it again;
}
```

The next time you want to use these same instructions, rather than writing out the individual steps, you can simply call the function named `buildFlatpackFurniture()`.

TIP

Although not compulsory, the Camel Case naming convention exists for function and variable names that contain multiple words. (If the convention were a function name, it would be written as CamelCase or camelCase.) Because these names can't have spaces, you need a way to distinguish where all the words start and end; otherwise, it takes a lot longer to scan them. The convention is to capitalize the first letter of each word after the first. This convention greatly improves the readability of your code, so I highly recommend that you adhere to this rule in all your sketches for your benefit and the benefit of those reading your code!

The word `void` is used for a function that returns no value, and the word that follows is the name of that function. In some circumstances, you might either put one or more values into a function or expect one or more values back from it, the same way you might put numbers into a calculation and expect a total back.

REMEMBER

You must include `void setup` and `void loop` in every Arduino sketch; they're the minimum required to upload. But it's also possible to write your own custom functions for the task you need to do. For now, just remember that you have to include `void setup` and `void loop` in every Arduino sketch you create. Without these functions, the sketch will not compile.

Setup

Setup is the first function an Arduino program reads, and it runs only once. Its purpose, as hinted in the name, is to set up the Arduino device, assigning values and properties to the board that do not change during its operation. In the Blink sketch, the `setup` function looks like this:

```
// the setup function runs once when you press reset or power the board
void setup() {
  // initialize digital pin LED_BUILTIN as an output.
  pinMode(LED_BUILTIN, OUTPUT);
}
```

TIP

Note on your screen that the word `void` is turquoise and the word `setup` is green. These colors indicate that the Arduino software recognizes these words as *core* functions, as opposed to a function you have written. If you change the case of the words to `Void Setup`, they turn black, which illustrates that the Arduino code is *case sensitive*. Having the correct case is an important point to remember, especially when it's late at night and the code doesn't seem to be working.

The contents of the `setup` function are contained within the curly brackets, `{` and `}`. Each function needs a matching set of curly brackets. If you have too many of either bracket, the code does not compile, and you are presented with an error message like the one in Figure 3-10.

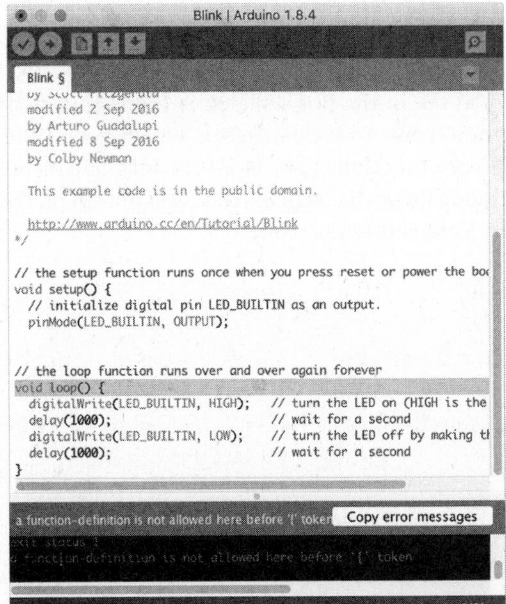

FIGURE 3-10: The Arduino software is telling you that a bracket is missing.

PinMode

The `pinMode` function configures a specified pin for either input (receive data) or output (send data). The function includes two parameters:

- `pin`: The number of the pin whose mode you want to set
- mode: Either `INPUT` or `OUTPUT`

In the Blink sketch, after the single-line comment, you see this line of code:

```
pinMode(LED_BUILTIN, OUTPUT);
```

The word `pinMode` is colored orange. As I mention previously in this chapter, highlighted text indicates that Arduino recognizes the word as a core function. `LED_BUILTIN` and `OUTPUT` are colored blue so that they can be identified as *predefined variables*, values that the Arduino integrated development environment (IDE) recognizes. In this case, `LED_BUILTIN` tells the function that you want to control the onboard LED on the Arduino board (marked *L*) and `OUTPUT` sets the mode of that pin. I go into more detail about variables in Chapter 6.

That's all you need for `setup`. Next up, the `loop` section.

Loop

The next section you see in the Blink sketch is the `loop` function, with the word `void` in turquoise and the word `loop` in green, which means the Arduino software recognizes it as a core function. `loop` is a function, but instead of running one time, it runs continuously until you press the reset button on the Arduino board or remove the power. Here is the `loop` code:

```
// the loop function runs over and over again forever
void loop() {
  digitalWrite(LED_BUILTIN, HIGH);   // turn the LED on (HIGH is the voltage level)
  delay(1000);                        // wait for a second
  digitalWrite(LED_BUILTIN, LOW);    // turn the LED off by making the voltage LOW
  delay(1000);                        // wait for a second
}
```

DigitalWrite

Within the `loop` function, you again see curly brackets and two orange functions: `digitalWrite` and `delay`.

First is digitalWrite:

```
digitalWrite(LED_BUILTIN, HIGH);   // turn the LED on (HIGH is the voltage level)
```

The comment says `turn the LED on`, but what exactly does that mean? The `digitalWrite` function sends a digital value to a pin. As mentioned in Chapter 2, digital pins have only two states: on or off. In electrical terms, these can be referred to as either a `HIGH` or `LOW` value, which is relative to the voltage of the board.

An Arduino Uno requires 5V to run, which is provided by either a USB or an external power supply (as mentioned in Chapter 2), which the Arduino board reduces to 5V. This voltage is interpreted by the Arduino. A `HIGH` value is equal to 5V and `LOW` is equal to 0V.

The `digitalWrite` function includes two parameters:

- `pin`: The name or number of the pin whose mode you want to set
- `value`: Either `HIGH` or `LOW`

So `digitalWrite(LED_BUILTIN, HIGH);` in plain English would be "send 5V to the onboard LED on the Arduino," which is enough voltage to turn on an LED.

Delay

In the middle of the `loop` code, you see this line:

```
delay(1000);                       // wait for a second
```

This function does just what it says: It stops the program, in this case for an amount of time in milliseconds. In this case, the value is 1000 milliseconds, which is equal to one second. During this time, nothing happens. Your Arduino is chilling out, waiting for the delay to finish.

The next line of the sketch provides another `digitalWrite` function to the same pin, but this time writing it low:

```
digitalWrite(LED_BUILTIN, LOW);    // turn the LED off by making the voltage LOW
```

This function tells Arduino to send 0V (ground) to the onboard LED, which turns off the LED. This line of code is followed by another delay that pauses the program for one second:

```
delay(1000);                       // wait for a second
```

At this point, the program returns to the start of the loop and repeats itself, ad infinitum.

So the loop is doing this:

» Sending 5v to the LED

» Waiting a second

» Sending 0v to the LED

» Waiting a second

As you can see, this gives you the blink!

Blinking Brighter

The LED marked L is actually connected just before it reaches pin 13. On earlier boards, it was necessary to provide your own LED. But because the LED proved so useful for debugging and signaling, one is now in permanent residence to help you out.

For this next bit, you need a loose LED from your kit. LEDs come in a variety of shapes, colors, and sizes but should look something like the one in Figure 3-11.

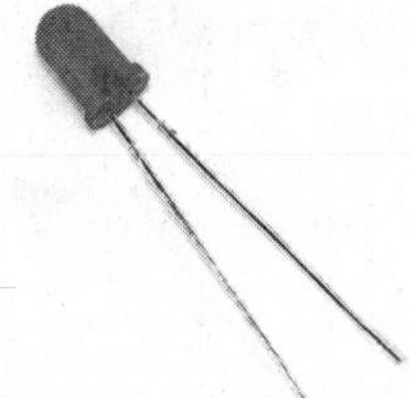

FIGURE 3-11: A lone LED, ready to be put to work.

Take a look at your LED and note that one leg is longer than the other. Place the long leg (anode or +) of the LED in pin 13 and the short leg (cathode or nd) in GND (ground). See Figure 3-12. You see the same blink, but it is (I hope) bigger and brighter depending on the LED you use.

FIGURE 3-12: Arduino LED pin 13.

As mentioned previously, the `LED_BUILTIN` variable is a predefined variable that refers to the onboard LED. Because the LED is controlled by pin 13 on your Arduino, it's possible to control a different component by connecting it to pin 13. The voltage supplied by all other pins can be too high for most LEDs. Fortunately, another feature of pin 13 is its built-in, pull-down resistor. This resistor keeps your LED at a comfortable voltage and ensures that it has a long and happy life.

Tweaking the Sketch

I've gone over this sketch in great detail, and I hope everything is making sense. The best way to understand what is going on, however, is to experiment! Try changing the delay times to see what results you get. Here are a couple of things you can try:

- Make the LED blink the SOS signal.
- See how fast you can make the LED blink before it appears to be on all the time.

While experimenting, it's wise to save your sketch under a different name. If the name is descriptive, such as sosLED, you can more easily find your project again. Also, each time you make a change to your code, remember to repeat the steps in this chapter to verify the code and upload it to your Arduino.

2 Getting Physical with Arduino

IN THIS PART . . .

Find out more about the prototyping tools you need to build your projects.

Dip into a bit of electronics theory.

Discover new and interesting things that your Arduino can do by building a few basic examples.

» Assembling a tool kit

» Becoming an electronics super sleuth, with a multimeter

Chapter 4
Tools of the Trade

In Chapter 3, I cover one of the most basic Arduino applications: blinking an LED. This application requires only an Arduino and a few lines of code. Although blinking an LED is fun, you can use an Arduino for an almost unlimited number of other things — making interactive installations, controlling your home appliances, and talking with the Internet, to name a few.

In this chapter, you branch out by gaining an understanding of prototyping and how to use some basic prototyping tools to do more with your Arduino. Prototyping tools allow you to make temporary circuits to try new components, test circuits, and build simple prototypes. In this chapter, you find out about all the equipment and techniques you need to build your own circuits and prototype your ideas.

Finding the Right Tools for the Job

Prototyping is all about exploring ideas, which also happens to be the core of what Arduino is all about. Although theory is important, you often learn better and faster from doing an experiment.

This section introduces you to some prototyping tools and components that you can use to start building circuits. You can then use these circuits to form the basis for your own projects.

A great number of tools are at your disposal to help you experiment. This section contains a short list of the recommended ones. Breadboards and jumper wires are included in most kits and are essential for building circuits for your Arduino project. Needle-nose pliers are not essential but are highly recommended.

Breadboard

Breadboards are the most essential part of your prototyping kit. They are the base on which you can prototype your circuits. Breadboards allow you to temporarily use components rather than committing them to a circuit and soldering them in place. (Soldering is covered in Chapter 9.).

Breadboards get their name from a practice used in the early 1900s. At that time, components were a lot bigger, and people would prototype circuits by fixing them to an actual breadboard — that is, a board intended for cutting bread on. The components were joined by wrapping lengths of wire around nails that came in contact with the components. By unwrapping the wire from one nail and wrapping it around another, you could quickly change the circuit.

Modern breadboards are much more refined. The outside consists of a plastic case with rows and columns of holes, underneath which are tracks of copper. These tracks allow you to quickly and easily connect components electrically.

Breadboards come in different shapes and sizes, and many have connectors to allow you to lay them out in the arrangement you need.

Figure 4-1 shows a fairly standard breadboard. If you were to remove the cream-colored plastic coating, two parallel copper tracks would run down each of the long sides of the board. These copper lengths are generally used to provide a source of power (PWR) and ground (GND) and are sometimes marked with a positive (+) or negative (−) symbol or a red and black or a red and blue line. Red is always positive (+) and black or blue is negative (−).

A *power rail*, or *ground rail*, is basically a source of voltage or ground. Circuits often need power or ground for a variety of different functions, and one source often isn't enough. When you have multiple wires that all need to get to the same place, you use a rail. From this rail, jump wires can source whatever is needed.

TIP

Although these tracks are marked, you can use them for anything. However, keeping to convention ensures that other people can easily understand your circuit, so I advise that you do so.

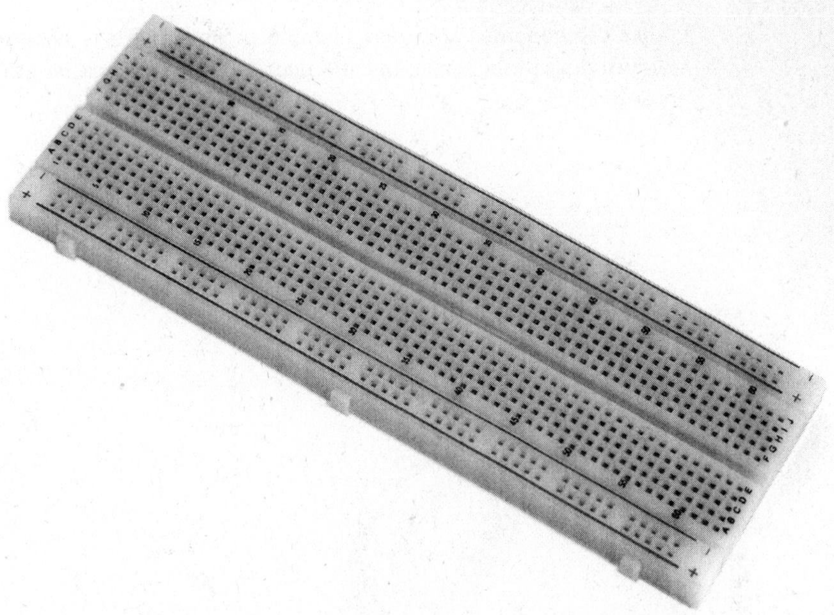

FIGURE 4-1:
A breadboard.

Down the middle of the breadboard are lots of short tracks running parallel to the short edge, separated with a trench down the middle. You use this middle trench to mount components and integrated circuits, including pushbuttons (discussed in Chapter 6) and optocouplers (see the bonus chapter "Hacking Other Hardware" at `www.dummies.com/go/arduinofd`). The trench makes it easier to lay out your circuit and also provides space for jump wires to connect your components to the places they need to get to.

When you place a jump wire or a component into one of the breadboard's sockets, you should feel a bit of friction. A pincer-like device holds the wire or component in place. It provides enough grip to hold things in place for a project while working at a desk, but it's loose enough for you to easily remove it with your fingers.

I have seen people taping breadboards into boxes for their projects, but this practice isn't advisable. Smaller components or loose wires can easily come loose, and it can be a pain to figure out why something isn't working. If you have a project working and want to get it out into the real world, jump ahead to soldering (explained in Chapter 9). Soldering makes your project last longer, and it's a lot of fun, too!

Jump wires

Jump wires (shown in Figure 4-2) are short lengths of insulated equipment wire. You use them to connect your components to the rows of your breadboard, other components, and your Arduino. A jump wire is no different from other wire in a material sense, but it's usually cut to a short length that is useful for breadboards.

FIGURE 4-2: A selection of jump wires.

You can find insulated wire everywhere. It can be the thick power cord used to plug in a household appliances or it can be much thinner, such as wire used for an earphone. Insulated wire is basically a conductive metal wire surrounded by an insulator that protects you from the electricity and protects the electrical signal from any outside interference.

The wire you use most often in this book (and most useful for your Arduino projects) could be subcategorized as *insulated equipment wire*. This type of wire is generally used on a small scale for low-voltage electrical applications.

The equipment wire you use is one of two varieties: single core and multicore. *single-core wire* is a single piece of wire, just as a coat hanger is a single piece. Single-core wire is extremely good for holding its shape but shears if you bend it too much. Therefore, this type of wire is useful for laying out wires on your breadboard neatly as long as you won't need to move them much.

Multicore wire can have the same diameter as single core but consists of lots of little wires instead of just one wire. The little wires are twisted together, giving the multicore wire more strength and resistance to bending than a single-core wire. This twisted design is the same technique used on suspension bridges. Multicore wires are especially suited to connections that will change often, and they last a long time.

You can cut jump wires yourself, but they also come in handy packs of assorted colors and lengths. Cutting them yourself is significantly cheaper because you can buy a large reel of wire. On the other hand, if you want a variety of colors, you have to buy a reel for each color, which can amount to a fairly big investment for your first circuit. The packs save you this initial cost and give you the variety you need. You can buy the large reels when you know you'll need them.

Also, when considering cutting your own wire, bear in mind the difference in finishing for homemade jump wires. Single-core wires are much the same whether you or someone else cuts them yourself, but they deteriorate quickly as they bend. Multicore wires last longer, but if they are cut from a reel, you're left with a few small wires on the ends that can easily fray in the same way as the end of a piece of string or thread does. You have to fiddle with them often, twisting with your thumb and forefinger between uses to ensure that the small wires stay as straight and rigid as possible.

Premade multicore jump wires are usually either soldered together into a single point or have a connecting pin soldered to the end of the strands of wire. This design ensures that the connecting end is as reliable as single core while giving you the flexibility of multicore.

Ideally, you should have a pack of premade multicore jump wires to get you going. These packs are the most versatile and long-lasting choice for your prototyping kit. Eventually, you'll want to have a variety of jump wires so that you can be prepared for any situation when building a circuit.

Needle-nose pliers

Needle-nosed pliers, shown in Figure 4-3, are the same as regular pliers but with a very fine point, ideal for picking up tiny components. Electronics can be an extremely fiddly business, and it's extremely easy to mangle the delicate legs of components when pushing them into a breadboard. These specialist pliers are not essential but add a little bit of finesse to building circuits.

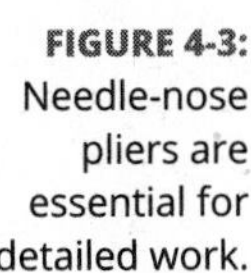

FIGURE 4-3: Needle-nose pliers are essential for detailed work.

Multimeter

A *multimeter* is a meter that measures volts, amps, and resistance. It can tell you the values of different components and what's going on with different parts of your circuit. Because you can't see what's going on in a circuit or component, a multimeter is essential for understanding its inner workings. Without it, you have to rely on guesswork, which is always a bad idea with electronics.

Figure 4-4 shows a good mid-range digital multimeter. As does this one, most multimeters include:

» **A digital display:** This display looks like the one on your digital alarm clock and shows the values you're reading. Because of the limited number of digits, the decimal place moves to accommodate a larger or smaller number. A multimeter with an auto-ranging function finds the number automatically. If your multimeter doesn't have this feature, you must change the mode manually to the range you require.

FIGURE 4-4: A good digital multimeter can often save your project.

» **A mode-selection dial:** This dial allows you to choose among the different functions on the multimeter. These functions can be for volts, amperes, and ohms as well as for the range within each of these, such as hundreds, thousands, tens of thousands, hundreds of thousands, and millions of ohms.

TIP

The best multimeters also include a continuity tester, which tells you whether your connections are actually connected by sounding a tone. Having a tone has saved me hours of work retracing my steps on projects, so I definitely recommend investing in a good mid-range multimeter with this feature.

- **A set of probes:** Probes (also called *test leads*) are the implements you use to test parts of your circuit. Multimeters come with two skewer-like probes that are designed to fit into tight places and are most useful for making contact with small components on circuit boards. You can also find probes with crocodile clips on the end, or you can simply buy crocodile clips that you can attach yourself. These clips are especially useful for grabbing onto wires.
- **A set of sockets:** The probes can be repositioned into different sockets depending on the use. In this case, the sockets are labeled A, mA, COM, and VΩHz. The socket marked A is for measuring large currents in amps (A), up to 20A. Note that the probe can read that current for only 10 seconds. These limits are indicated on a line between this socket and the COM socket, which also indicates that the two probes should be placed in A (red probe) and COM (black probe). The mA socket (red probe) is for smaller currents that are less than 500mA. The COM socket (black probe) is short for Common and is a point of reference for your measurements. In most cases, COM is the ground of your circuit and uses the black probe. The socket marked VΩHz (red probe) is used for measuring the voltage (V or volts), resistance (Ω or ohms), and frequency (Hz or hertz) between this socket and the COM port (black probe).

Using a Multimeter to Measure Voltage, Current, and Resistance

All Arduinists should know a few basic techniques for checking their circuits. Volts, amps, and current can all be calculated in advance (as you learn in Chapter 5), but in the real world, many other factors can arise that you can't account for. If you have a broken component or a faulty connection, you can lose hours guessing what could be wrong, so a multimeter becomes essential for solving problems in your circuit. In this section, you learn about measuring voltage, current, and resistance and checking the continuity of your connections.

Measuring voltage in a circuit

Measuring the voltage is essential, whether you're checking the voltage of a battery or the voltage passing through a component. If things aren't lighting up or whirring, a connection might be loose or you might have sent too much power and burnt out what you were trying to power. When something isn't working, it's time to check the voltage in your circuit and make sure it is correct.

First you need to check that the probes of your multimeter are in the correct sockets. Use the red probe in the socket marked V, for volts, and use the black probe in the socket marked COM, for ground. Next, set your multimeter to volts by using the dial in the center of the multimeter.

Some multimeters, such as the one in Figure 4-4, have a button to toggle between DC and AC. On other multimeters, a DC voltage is signified by a V followed by a square-shaped digital wave, as opposed to an AC voltage, which is indicated by a smooth, analog wave.

TECHNICAL STUFF

Voltage is measured in parallel. To measure in parallel you must bridge the part of the circuit that you want to measure without interfering in it. Figure 4-5 shows how you do this. The positive probe should always be on the positive side of the component and the negative on the other side. Getting the probes the wrong way won't cause any damage to your circuit but will give you a negative reading rather than a positive one.

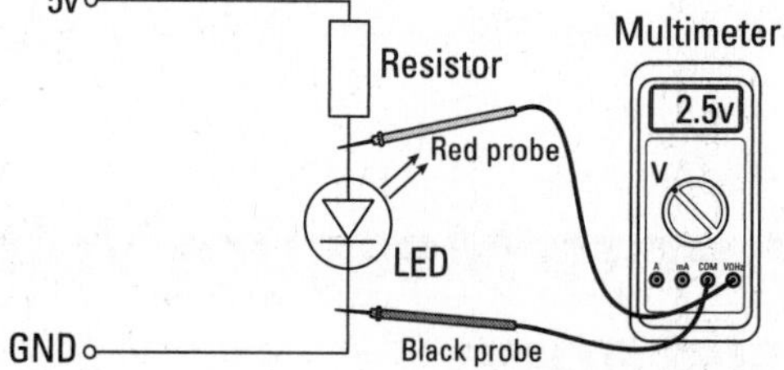

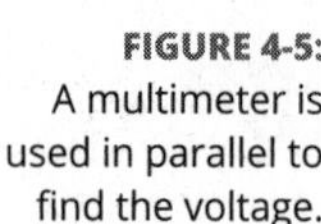

FIGURE 4-5: A multimeter is used in parallel to find the voltage.

A good way to test your multimeter is to measure the voltage between the 5V pin and GND on your Arduino. Make sure that your Arduino is plugged in, and connect a jump wire to each pin to make them easier to access with your probes. Place the red voltage probe on the 5V wire and the black common probe on the GND wire. Doing so should return a value of 5V on the multimeter screen and prove that your Arduino is supplying 5 volts as expected.

Measuring current in a circuit

You may have the right voltage, but there isn't enough current to power the light or motor you're driving. The best way to find out is to check the circuit to see how much current is being drawn and compare that to the power supply you're using.

Check that your probes are connected to the correct sockets of the multimeter. Some meters have two possible sockets, one for very high currents measured in amps (or A) and another for low currents measured in milliamps (mA). Most basic Arduino circuits require only a reading in milliamps, but circuits with large lights, motors, or other devices, require a reading in amps. Then turn the dial on your meter to select the correct level of amps, A or mA or even μA (microamps).

Current is measured in series, which means that the multimeter must be placed in line with the other components in the circuit so that the current flows through the multimeter as if it were another component. Figure 4-6 shows this series measurement in action.

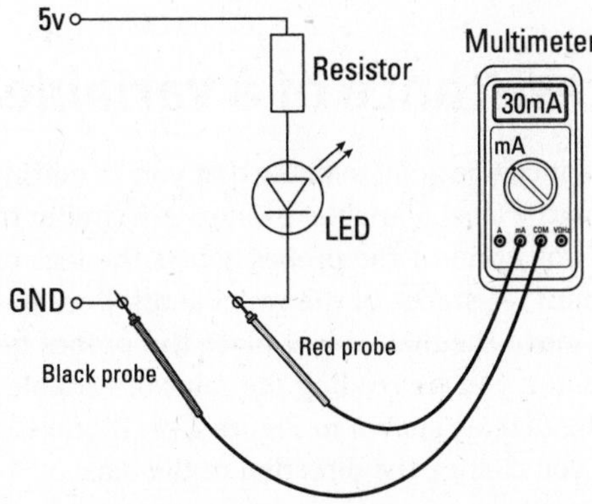

FIGURE 4-6: A multimeter is used in series to find the current.

By using the 5V and GND pins on your Arduino, you can use jump wires to build the circuit in Figure 4-6. By using your multimeter to connect the LED and GND, you will be able to read the current in the circuit. You'll learn how to build circuits similar to Figure 4-6 in Chapter 6.

TECHNICAL STUFF

If you're blinking or fading the output, your current changes, so you may want to set it to always be on to be sure of the maximum continuous current.

Measuring resistance of a resistor

Sometimes it can be difficult to read the value of a resistor, and it is necessary, or just easier, to confirm it with a multimeter. Simply set the multimeter to ohms or Ω and place one probe on each leg of the resistor, as shown in Figure 4-7.

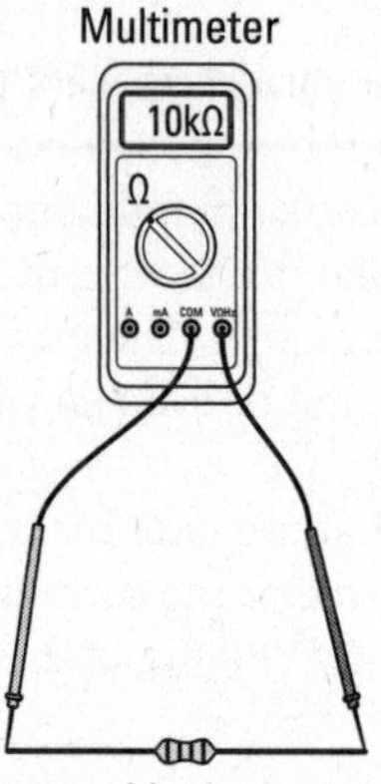

FIGURE 4-7: Finding the resistance of a resistor.

Measuring resistance of a variable resistor

With variable resistors, it can be good to know that you're getting the full range of resistances promised on the label. Variable resistors are similar to passive resistors but have three legs. If you connect the probes across the legs on either side, you are reading the maximum resistance of the variable resistor, and the reading does not change when you move the dial. If you place the probes between the center and one side of the resistor, you are reading the value of variable resistance, which changes as you turn the dial, as shown in Figure 4-8. If you switch to the center and the opposite side, you change the direction of the dial.

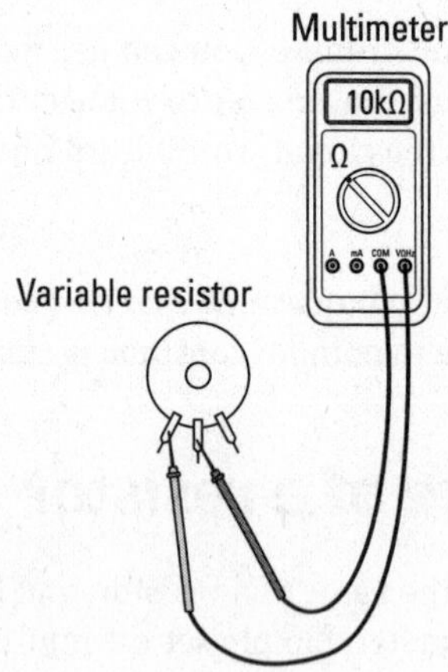

FIGURE 4-8: Finding the resistance of a variable resistor.

Checking the continuity (in bleeps) of your circuit

If you have a high-quality multimeter, it should have a continuity tester, represented by a speaker or sound symbol on the dial. You use the continuity tester to verify that parts of your circuit are connected; the multimeter communicates the connection by bleeping, ideally producing a continuous tone when the connection is good.

First test the continuity feature itself by turning your dial to the continuity test symbol and touching the ends together. If you hear an unbroken tone, the feature is working. Next, test the connection in your circuit by placing the probes along any length of wire or connection, as shown in Figure 4-9.

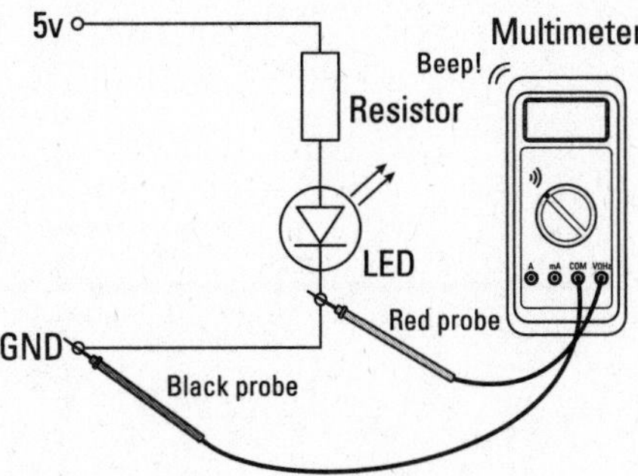

FIGURE 4-9: Checking the continuity of your circuit.

IN THIS CHAPTER

» Coming to grips with electricity

» Understanding a few useful equations

» Finding your way around circuit diagrams

» Realizing the importance of color coding

Chapter 5

A Primer on Electricity and Circuitry

In this chapter, you look at the fundamentals of electricity. In later chapters, you delve deeper into electronics, so it's important that you have a basic understanding of how electricity behaves in your circuit.

The great thing about Arduino is that you don't need to study electronics for years to use it. That said, it's a good idea to know a bit of the theory to back up the practical side of your project. In this chapter, you look at a few equations to help you build a balanced and efficient circuit, you look at circuit diagrams, which provide you with a roadmap of your circuit, and you learn a bit about color coding, which can make your life easier when building circuits.

Understanding Electricity

Most people take electricity for granted but find it difficult to define. Simply put, *electricity* is a form of energy resulting from the existence of charged particles (such as electrons or protons), either statically as an accumulation of charge or dynamically as a current.

This definition of electricity is describing electricity on an atomic level, which is more complex than you need to know when dealing with the circuitry in your Arduino projects. Your main concern is simply to understand that electricity is energy and has a current. For those of you who want to understand electricity at this level, you can check out *Electronics For Dummies*, 3rd Edition, by Cathleen Shamieh.

To illustrate the idea of the flow of a current, take a look at a simple light switch circuit (see Figure 5-1). The circuit is similar to those you may have made in physics or electronics classes at school with no Arduino involved, using only a battery, a switch, a resistor, and an LED.

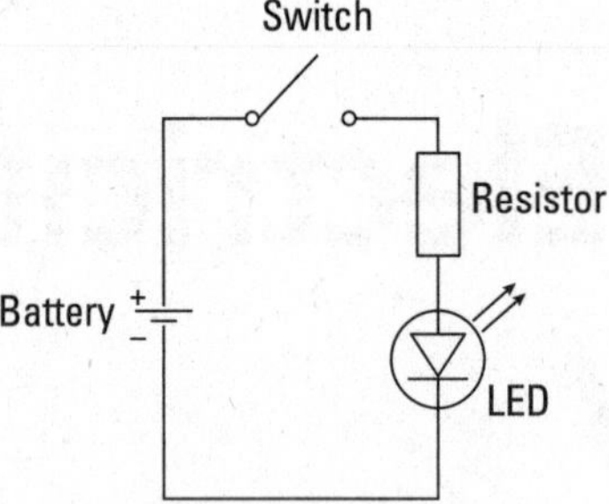

FIGURE 5-1: A basic light switch circuit.

In this circuit, you have a source of electrical power in the form of a battery. Power is supplied in watts and is made up of voltage (in volts) and current (in amps). Voltage and current are supplied to the circuit through the positive (+) end of the battery.

You use a switch to control the power to the circuit. The switch can either be open, which breaks the circuit and stops the flow of current, or closed, which completes the circuit and allows it to function.

The power can be used for various applications. In this case, the circuit is powering an LED. The battery is supplying the LED with 4.5V, which is more than the LED needs to light. If the LED were supplied with this much voltage, you would risk damaging it, so you need a resistor before it to resist the voltage. Also, if the voltage is too low, the LED will not reach full brightness.

To complete the circuit, the power must return to ground at the negative (−) end of the battery. The LED draws as much current as necessary to light to full brightness.

By drawing current, the LED is also resisting the flow, ensuring that only the required current is drawn. Unless the current is used or resisted by components, the circuit draws all the available current as quickly as possible. This is known as a *short circuit*. An example of a short circuit is connecting the positive directly to the negative, with no components in between.

The basic principles that you need to understand follow:

- » An electrical circuit is, as the name suggests, a circular system.
- » The circuit needs to use the power inside it before it returns to the source.
- » If the circuit does not use the power, that power has nowhere to go and can damage the circuit.
- » The easiest way to interact with a circuit is to break it. By controlling when and where the power is, you have instant control over the outputs.

Using Equations to Build Your Circuits

You are now aware of a few characteristics of electricity:

- » Power (P) is in watts, such as 60W
- » Voltage (V) is in volts, such as 12V
- » Current (I) is in amps, or amperes, such as 3A
- » Resistance (R) is in ohms, such as 150Ω

These characteristics can be quantified and put into equations, which allow you to carefully balance your circuit to ensure that everything works in harmony. A variety of equations exist for determining all manner of attributes, but in this section I cover two basic ones that will be the most useful to you when working with Arduino: Ohm's Law and Joule's Law.

Ohm's Law

Perhaps the most important relationship to understand is that among voltage, current, and resistance. In 1827, Georg Simon Ohm discovered that voltage and current were directly proportional if applied to a simple equation (recall from the preceding list that *I* stands for current, measured in *amps*, or *amperes*):

$$V = I \times R$$

This equation became known as *Ohm's Law*. Using algebra, the equation can be rearranged to give you any one value from the remaining two:

$$I = V / R$$

or

$$R = V / I$$

You can take a look at this equation at work in an actual circuit. Figure 5-2 shows a simple circuit with a power source and a resistor.

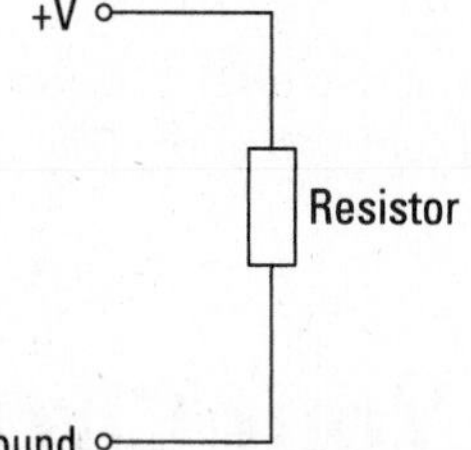

FIGURE 5-2: A power source and resistor.

In many situations, you will know the voltage of your power supply and the value of the resistor, so you can first calculate the current of the circuit as follows:

$$I = V / R = 4.5V / 150\Omega = 0.03A$$

This equation works in any order you want to put the values:

$$R = V / I = 4.5V / 0.03A = 150\Omega$$
$$V = I \times R = 0.03A \times 150\Omega = 4.5V$$

The easiest way to remember Ohm's Law is as a pyramid (see Figure 5-3). By eliminating any one element from the pyramid, you are left with the equation.

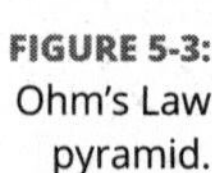

FIGURE 5-3: Ohm's Law pyramid.

"But how is this calculation useful to me in the context of Arduino?" I hear you cry. Here's a practical example that you might run into in a basic Arduino circuit.

The digital pins on an Arduino can supply up to 5V, so this is the most common supply of power that you use. An LED is one of the most basic outputs you want to control, and a fairly standard LED requires a voltage of 2V and about 30 milliamps or 30mA (0.03A) of current.

If you were to plug in the LED directly to the power supply, you would promptly witness a bright light followed by a plume of smoke and a burning smell. You're not likely to want that! To make sure that you can use the LED again and again safely, you should add a resistor.

Ohm's Law tells you that

$$R = V / I$$

But you also have to include two different voltage values, the voltage of the power supply (supply voltage) and the voltage required to power the LED (forward voltage). *Forward voltage,* a term that is often found in datasheets, especially when referring to diodes, indicates the recommended amount of voltage that the component can take in the direction that the current is intended to flow. For LEDs, this direction is from anode to cathode, with the anode connected to positive and the cathode to negative.

When referring to non-light-emitting diodes (covered in Chapter 7), you're using them to resist the flow of current in the opposite direction, from cathode to anode. In this case, the term is *reverse voltage*, which indicates the value in volts that the circuit must exceed for current to flow through the diode.

In this case, the voltages are labelled V_{SUPPLY} and $V_{FORWARD,}$ respectively. The Ohm's Law equation requires the voltage across the resistor (voltage that passes through the resistor), which is equal to the supply voltage minus the LED forward voltage, or

$$V_{SUPPLY} - V_{FORWARD}$$

The new equation looks like this:

$$R = (V_{SUPPLY} - V_{FORWARD}) / I = (5V - 2V) / 0.03A = 100\Omega$$

This tells you that you need a 100-ohm resistor to power an LED safely; the example circuit is shown in Figure 5-4.

Calculating power

To calculate the power consumption of your circuit in watts, you multiply the voltage and current of the circuit. The equation is

$$P = V \times I$$

If you apply this equation to the same circuit as the example in the "Ohm's Law" section, earlier in this chapter, you can calculate its power:

$$P = (V_{SUPPLY} - V_{FORWARD}) \times I = (5V - 2V) \times 0.03A = 0.09W$$

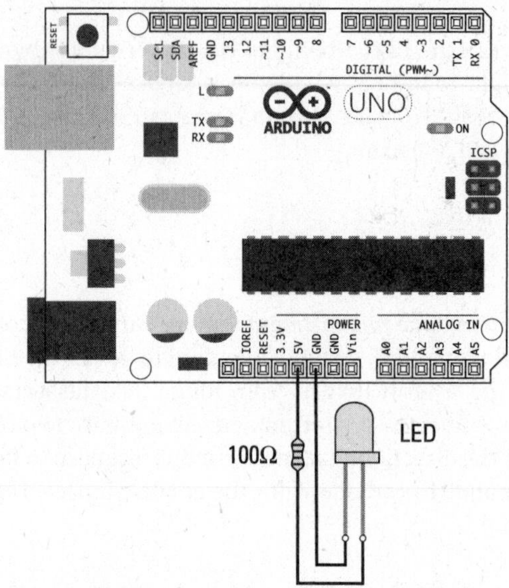

FIGURE 5-4: Applying Ohm's Law to an Arduino circuit.

This algebra works in the same way as Ohm's Law and can be reconfigured to find the missing value:

$$V = P / I$$

$$I = P / V$$

This calculation is useful because some types of hardware, such as a light bulb, show only the power and voltage rating, leaving you to figure out the current draw. The calculation is especially useful if you are trying (or failing) to run power-hungry devices, such as lighting or motors, off your Arduino pins. A USB-powered Arduino is capable of supplying 500mA of current, but an Arduino Uno can supply only a maximum of 40mA per pin *and* 200mA total from all the pins being used, which is not much at all. (You can find more details at `http://playground.arduino.cc/Main/ArduinoPinCurrentLimitations.`).

This calculation is simple, but you can combine it with your knowledge of Ohm's Law to fill in the blanks in a number of different circuits.

Joule's Law

Another man who gave his name to an equation was James Prescott Joule. Although not as well known as Georg Simon Ohm, he discovered a similar and perhaps complementary mathematical relationship between power, current, and resistance in a circuit.

Joule's Law is written like so:

$$P = I^2R$$

The best way to understand it is to look at how you arrive at it.

$$\begin{aligned} &\textit{If}, \quad V = I \times R\,(\text{Ohm's Law}) \text{ and } P = I \times R\,(\text{Power Calculation}) \\ &\textit{Then}, \quad P = I \times (I \times R) \end{aligned}$$

Which can also be written as

$$P = I^2 \times R$$

If this equation is applied to the circuit in Figure 5-2, you can discover the power consumption of the circuit:

$$P = I^2 \times R = (0.03A \times 0.03A) \times 100\Omega = 0.09W$$

This calculation tallies with our previous power calculation. You can calculate the power knowing only the current and resistance, or any combination of the three (power, current, and resistance).

$$\begin{aligned} &I = P / (I \times R) \\ &R = P / I^2 \end{aligned}$$

You can also do the same calculation for situations where you know only the voltage and resistance:

$$\begin{aligned} &\textit{If}, \quad I = V / R(\text{Ohm's Law}) \text{ and } P = I / V(\text{Power Calculation}) \\ &\textit{Then}, \quad P = (V / R) * V \end{aligned}$$

Which can also be written as

$$P = V^2 / R$$

Try the same circuit again to check the results:

$$((5V - 2V) * (5V - 2V)) / 100\Omega = 0.09W$$

This equation can also be rearranged into any combination, depending on which values you know:

$$\begin{aligned} &V = P / (V * R) \\ &R = V^2 / P \end{aligned}$$

Many Arduinists, myself included, are more practical than theoretical, attempting to build the circuit based on examples and documentation before doing the sums. This approach is perfectly all right and in the spirit of Arduino! In most cases, your circuit will have the desired outcome. However, with these few equations, it's possible to fill in the blanks in most circuits and ensure that everything is in order.

Working with Circuit Diagrams

Recreating circuits from photos or illustrations can be difficult, and for that reason, standardized symbols are used to represent the variety of components and connections in a circuit. These *circuit diagrams* are like maps of the subway system: They show you every connection clearly but have little resemblance to the way things look or connect in the physical world. The following sections delve a bit more into circuit diagrams.

A simple circuit diagram

This section looks at a basic light switch circuit made up of four components: a battery, a pushbutton, a resistor, and an LED, as shown in Figure 5-5.

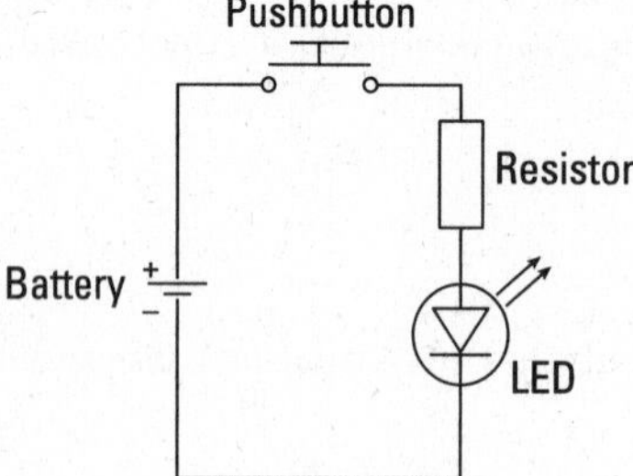

FIGURE 5-5: A simple light switch circuit diagram.

Table 5-1 shows the individual symbols for each component.

TABLE 5-1 **Basic Symbols**

Name	Symbol
Battery	
Pushbutton	
Resistor	
LED	

Figure 5-6 shows the same circuit laid out on a breadboard. The first thing you may notice is that this example has no battery. Because your Arduino has a 5V pin and a GND pin, these take the place of the positive (+) and negative (–) of the battery and allow you make the same circuit. The second thing you may notice is that the circuit uses a pushbutton and, therefore, is not technically a light *switch*. Many pushbuttons are made to fit a breadboard, which is more convenient.

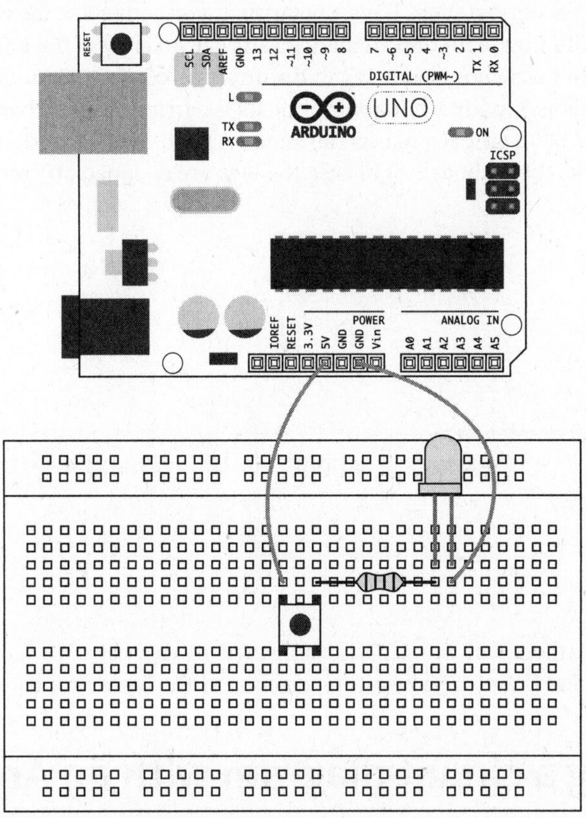

FIGURE 5-6: A simple light switch circuit laid out on a breadboard.

I find that the best way to compare a circuit diagram to the actual circuit is to follow the connections from positive to negative.

If you start at the positive (+) 5V pin on the Arduino, it leads to the pushbutton. The physical pushbutton has four legs, whereas the symbol has only two. The legs of the physical pushbutton are mirrored so that two are connected on one side and two on the other. For this reason, it's important to get the orientation of the pushbutton right. The physical switch has four legs to make it more versatile, but as far as the circuit diagram is concerned, there is only one switch with one line in and one line out.

The other side of the pushbutton is connected to a resistor. Other than the fact that the resistor symbol in the diagram is not as bulbous as the physical resistor, the diagram and physical resistor match up well; one wire goes into the resistor and another goes out. The value of the resistor is written alongside the component, as opposed to the color-coded stripes on the physical one. Resistors do not have polarity (no positive or negative), so there is nothing else to show.

An LED, by contrast, does have a polarity. If you connect it the wrong way around, it won't illuminate. In the circuit diagram, the polarity of the LED is marked by the triangle in the symbol, pointing in the direction of the current flow from + (anode) to – (cathode), with a horizontal line as a barrier in the other direction. On the physical LED, a long leg marks the anode (+) and the flat section on the side of the lens marks the cathode (–),in case the legs are chopped off. See Figure 5-7.

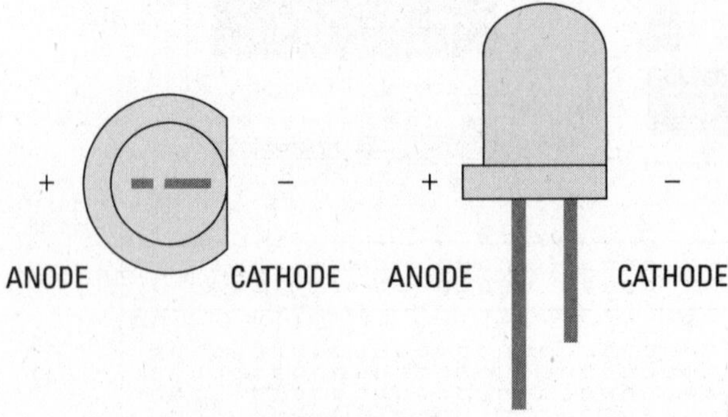

FIGURE 5-7: A long leg or a flat side marks an LED's polarity.

The – (cathode) of the LED is connected to the negative (–) GND pin on the Arduino and then to the negative terminal of the battery to complete the circuit.

Using a circuit diagram with an Arduino

Although it's useful to understand the simple circuit in Figure 5-6, you will most likely be using an Arduino in your circuit somewhere, so take a look again at the same circuit powered from an Arduino (see Figure 5-8).

This circuit has more components than the circuit described in the previous section.

The Arduino is on the left in the diagram. This symbol is standard for an integrated circuit and is similar to its physical representation — a rectangle with lots of legs poking out. All the legs or pins are labeled so that you can tell them apart.

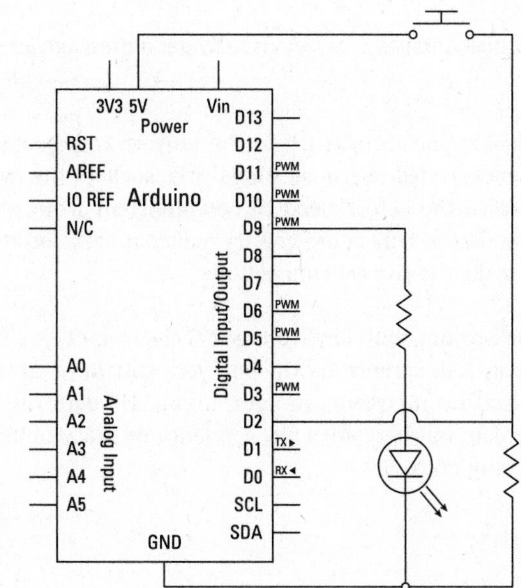

FIGURE 5-8: A light switch circuit running through an Arduino.

Also, rather than have one circuit, the diagram shows two, each running back to the Arduino, to illustrate how the Arduino fits in with conventional circuits. Instead of switching the power on and off, you're sending a signal to the Arduino, which interprets it and outputs it to the LED.

Here is a great practice to adopt when you have a complicated circuit: Rather than tackle it as a whole, break it up into its components. This circuit has one input circuit and one output. I describe this circuit in more depth in Chapter 6.

Color-Coding

An important technique in electronics is color-coding, and it becomes even more important as you progress to more complex circuits. Wiring circuits correctly can be hard enough, but staring at a sea of same-colored wires makes the task infinitely harder.

You're probably aware of color-coding even if you're new to electronics. Traffic lights, for example, are color coded to give drivers a clear message of what to do:

- Green means proceed.
- Amber means prepare to stop.
- Red means stop.

Color-coding is a quick and easy way to visually get a message across without lots of words.

All sorts of electrical applications, such as the 120v or 240v plugs and sockets in your home, are color-coded. Because plugs and sockets are widely used and potentially dangerous, the colors need to be consistent from plug to plug and match national standards. This color-coding makes it easy for any electrician or DIY enthusiast to make the correct connections.

Because you will be working with low-voltage DC electronics, you have less potential for causing yourself serious harm, but you still have great potential for destroying the delicate components in your circuit. No definitive rules exist for organizing your circuit, but here are a few conventions that can help you and others know what's going on:

- » Red is positive (+).
- » Black is negative (-).
- » Different colors are used for different signal pins.

These conventions are true on most breadboard circuits. Power and ground colors can change; for example, they can be white (+) and black (−) or brown (+) and blue (−), sometimes depending on the wire that is available to the person. As long as you use a color-coding system of some sort (and it's consistent), reading, copying, and fixing the circuit will be easier.

I've fixed many circuits that were broken because of the simple error of connecting the wires to the wrong places.

If the color-coding of the wire is ever questionable, check the connection (using the continuity checker on your multimeter) or the voltage running through the wire (using the voltage meter on your multimeter) to make sure that everything is as expected.

Datasheets

Picture the scene. Your friend has heard that you know a bit of electronics and has asked you to look at a circuit, copied from the Internet, that isn't working. But the board has lots of nondescript integrated circuits, so what do you do? The answer: Google!

The world contains millions, if not billions, of different components. The information you need to make sense of them is normally presented in the form of a *datasheet*. Every component should have a datasheet, provided by the manufacturer, that lists every detail of the component (often more detail than you need).

The easiest way to find a datasheet is to Google it. To find the right one, you need to know as much about the component as you can find out. The most important information for search purposes is the model number of the component. Figure 5-9 shows the model number of a transistor: P2N2 222A B01. If you Google that number plus the word *datasheet*, you should locate numerous PDF files that provide details about the component. If you can't find a model or part number, try to find it out from the place where you purchased the component.

FIGURE 5-9: The small print on this transistor tells you exactly what it is.

Resistor Color Charts

Resistors are extremely important to Arduino projects, and you can find a great variety of them to allow you to finely tune your circuit. Resistors can also be extremely small, making it impossible to write the resistance value on the resistor. For this reason, a color chart system exists to tell you what you need to know about these tiny components. If you take a close look at a resistor, such as the one in Figure 5-10, you can see a number of colored bands, which indicate the value in ohms of the resistor.

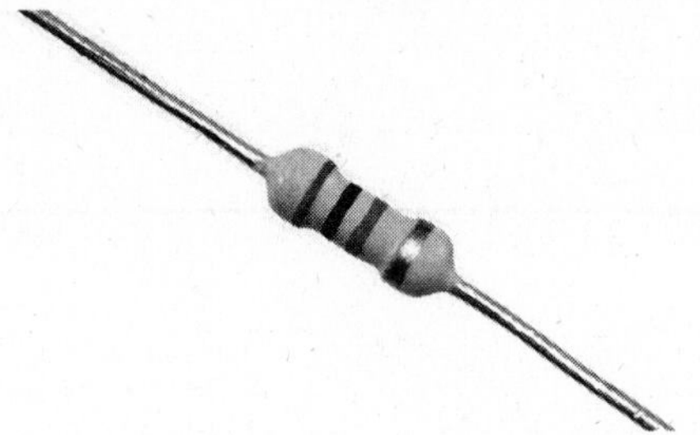

FIGURE 5-10: The color bands on a resistor.

A standard resistor has four bands that represent the following:

- First digit
- Second digit
- Multiplier
- Tolerance

TIP

Resistors are small and some colors can be difficult to distinguish unless you have extremely good vision. You can use a magnifying glass or smartphone camera to see the colors of the bands more clearly.

Table 5-2 lists the value in ohms, the multiplier, and the tolerance value that each color represents. First, you need to know the order in which to read the bands. Normally, you will see an equal-sized gap between the first three bands and a larger gap separating the fourth tolerance band.

TABLE 5-2 **Resistor Color Chart**

Color	Value	Multiplier	Tolerance
Black	0	$x10^0$	–
Brown	1	$x10^1$	±1%
Red	2	$x10^2$	±2%
Orange	3	$x10^3$	–
Yellow	4	$x10^4$	±5%
Green	5	$x10^5$	±0.5%
Blue	6	$x10^6$	±0.25%
Violet	7	$x10^7$	±0.1%
Gray	8	$x10^8$	±0.05%
White	9	$x10^9$	–
Gold	–	$x10^{-1}$	±5%
Silver	–	$x10^{-2}$	±10%
None	–	–	±20%

For example, here a few values you might find in your kit:

> Orange (3), Orange (3), Brown (x10^1), Gold $(\pm5\%) = 33 * 10 = 330$ ohms with ±5% tolerance
>
> Red (2), Red (2), Red (x10^2), Gold $(\pm5\%) = 22 * 10 * 10 = 2.2\text{K}$ ohms ±5% tolerance
>
> Brown (1), Black (0), Orange (x10^3), Gold $(\pm5\%) = 10 * 10 * 10 * 10 = 10\text{K}$ ohms ±5% tolerance

Seeing the colors and sometimes even telling which end to start reading from can be difficult. So, in most situations, it's best to use a multimeter to check the value of your resistor in ohms. (Remember to set the dial on your multimeter to ohm, or Ω.)

TIP

Resistors of the same value are often supplied on a reel of paper tape that holds the resistors together in a kind of ladder. This arrangement enables machines to feed in a reel of resistors in an orderly fashion before placing them on a PCB. If you write the value of that reel of resistors, you won't have to spend time reading or measuring the resistors each time you use them.

» Understanding inputs

» Varying resistances with potentiometers

» Showing off your stats with the serial monitor

Chapter 6

Basic Sketches: Inputs, Outputs, and Communication

In this chapter, I discuss some of the basic sketches needed to get you on your Arduino feet. This chapter covers a broad range of inputs and outputs using the sensors in your kit. If you don't yet have a kit, I suggest reading through Chapter 2 to find one of the recommended ones.

The Blink sketch (described in Chapter 3) gives you the basis of an Arduino sketch. This chapter expands it by adding circuits to your Arduino. I walk you through building circuits using a breadboard, as mentioned in Chapter 4, and introduce a few new components from your kit.

I detail uploading the appropriate code to your Arduino, walk you through each sketch line by line, and suggest tweaking the code yourself to gain a better understanding of it.

Uploading a Sketch

Throughout this chapter and much of the book, you learn about a variety of circuits, each with their respective sketches. The content of the circuits and sketches can vary greatly and are detailed in each of the examples in this book. Before you get started, you need to know one simple process for uploading a sketch to an Arduino board.

Follow these steps to upload your sketch:

1. **Connect your Arduino using the USB cable.**

   The square end of the USB cable connects to your Arduino and the flat end connects to a USB port on your computer.

2. **Select your board type and serial port.**

   If you're using the Web IDE, it automatically detects the board type and port. You can find the full list of supported boards by clicking the board's drop-down menu and clicking Select Other Board and Port.

   Using the downloadable IDE requires a few steps. Select your board type by choosing Tools ➪ Board ➪ Arduino/Genuino Uno from the menu bar. You can find a list of all available serial ports by choosing Tools ➪ Port ➪ com*X* or /dev/tty.usbmodem*XXXXX*, where *X* is a sequentially or randomly assigned number.

   In Windows, if you have just connected your Arduino, the COM port will normally be the highest number, such as com 3 or com 15. Many devices can be listed on the COM port list, and if you plug in multiple Arduinos, each one will be assigned a new number.

   On macOS, the /dev/cu.usbmodem *number* will be randomly assigned and can vary in length, such as /dev/cu.usbmodem1421 or /dev/cu.usbmodem262471. Unless you have another Arduino connected, it should be the only one visible.

3. **Click the Verify button.**

   The Verify button is a check mark. Your code is checked to make sure that it can be understood by the IDE. Common mistakes such as typos will be highlighted.

4. **Click the Upload button (arrow).**

   The code is compiled and sent to the Arduino.

Now that you know how to upload a sketch, you should be suitably hungry for some more Arduino sketches. To help you understand the first sketch in this chapter, I first tell you about a technique called pulse-width modulation (PWM). The next section briefly describes PWM and prepares you for fading an LED.

Using Pulse-Width Modulation (PWM)

In Chapter 2, I mention that sending an analog value uses something called pulse-width modulation (PWM). This technique allows your Arduino, which is a digital device, to act like an analog device. In the following example, PWM allows you to fade an LED rather than just turn it on or off.

Here's how it works: A digital output is either on or off. But it can be turned on and off extremely quickly thanks in part to the miracle of silicon chips. If the output is on half the time and off half the time, it is described as having a 50 percent duty cycle. The *duty cycle* is the period of time during which the output is active, so that could be any percentage — 20 percent, 30 percent, 40 percent, and so on.

When you're using LEDs as an output, the duty cycle has a special effect. Because the LED is blinking faster than the human eye can perceive, an LED with a 50 percent duty cycle looks as though it is at half brightness. This same effect allows you to perceive still images shown at 24 frames per second (or above) as a moving image.

With a DC motor as an output, a 50 percent duty cycle has the effect of moving the motor at half speed. In this case, PWM allows you to control the speed of a motor by pulsing it at an extremely fast rate.

So despite PWM's being a digital function, it is referred to as `analogWrite` because of the perceived effect it has on components.

The LED Fade Sketch

In this sketch, you make an LED fade on and off. In contrast to the sketch that resulted in a blinking LED in Chapter 3, you need some extra hardware to make the LED fade on and off.

For this project you need:

- An Arduino Uno
- A breadboard
- An LED
- A resistor (greater than 120 ohm)
- Jump wires

WARNING

It's always important to make sure that your circuit is not powered while you're making changes to it. You can easily make incorrect connections, potentially damaging the components. So before you begin, make sure that the Arduino is unplugged from your computer or any external power supply.

Lay out the circuit as shown in Figure 6-1, which is a simple circuit like the one used for the Blink sketch in Chapter 3 but uses pin 9 instead of pin 13. Pin 9, unlike pin 13, is capable of pulse-width modulation, which is necessary to fade the LED. However, note that pin 9 requires a resistor to limit the amount of current supplied to the LED. (On pin 13, this resistor is included on the Arduino board itself.)

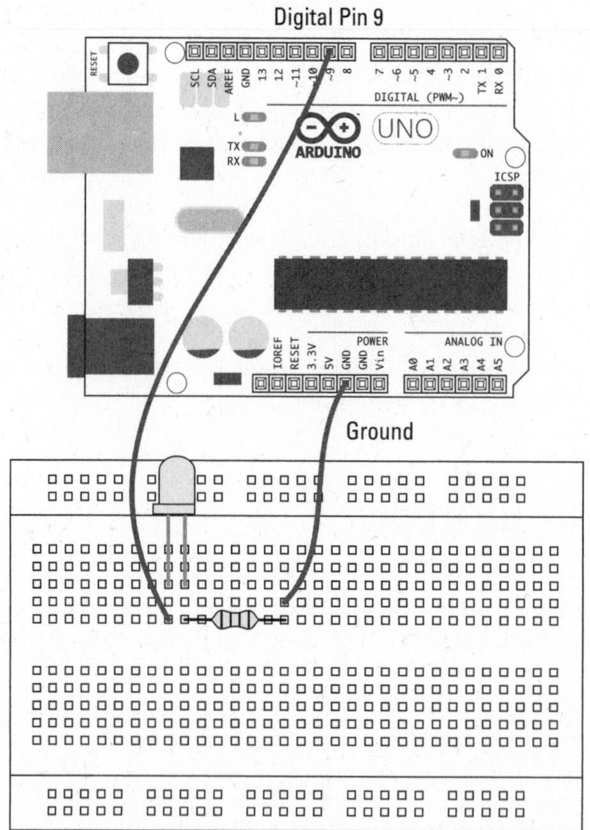

FIGURE 6-1: Pin 9 is connected to a resistor and an LED and then goes back to ground.

The schematic in Figure 6-2 shows you the simple circuit connection. The digital pin, pin 9, is connected to the long leg of the LED; the short leg connects to the resistor, which goes on to ground, GND. In this circuit, the resistor can be either before or after the LED, as long as it is in the circuit.

PUTTING UP RESISTANCE

As you learn in Chapter 5, calculating the correct resistance is important for a safe and long-lasting circuit. In the circuit you're building in this chapter, you are potentially supplying your LED with a source of 5V (volts), the maximum that a digital pin can supply. A typical LED such as those in your kit has an approximate maximum forward voltage of 2.1V (volts), so a resistor is needed to protect it. The LED draws a maximum current of approximately 25mA (milliamps). Using these figures, you can calculate the resistance (ohms):

$$R = (V_S - V_L) / I$$
$$R = (5 - 2.1) / 0.025 = 116 \text{ ohms}$$

The nearest fixed resistor above this calculation that you can buy is 120 ohms (brown, red, brown), so you're in luck if you have one of those. If not, you can apply the rule of using the nearest resistor above this value. This resistor resists more voltage than the optimum, but your LED is safe and you can always switch out the resistor later when you want to make your project more permanent. Suitable values from various kits include 220Ω, 330Ω, and 560Ω.

You can always refer to Chapter 5 to find your resistor value on the color chart or use a multimeter to measure the value of your resistors. There are even smartphone apps that have resistor color charts (although this may be a source of great embarrassment and ridicule among friends).

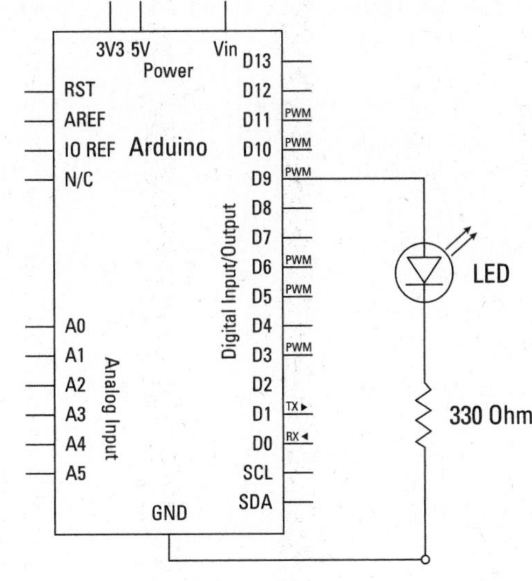

FIGURE 6-2: A schematic of the circuit to fade an LED.

TIP

It's always a good idea to *color code* your circuits — that is, use various colors to distinguish one type of circuit from another. Doing so greatly helps keep things clear and can make problem solving much easier.

You should adhere to a few good standards. The most important areas to color code are power and ground. These are nearly always colored red and black, respectively, but you might occasionally see them as white and black as well, as mentioned in Chapter 5. The other type of connection is usually referred to as a *signal wire*, which is a wire that sends or receives an electrical signal between the Arduino and a component. Signal wires can be any color that is not the same as the power or ground color.

After you assemble your circuit, you need the appropriate software to use it. From the Arduino menu, choose File⇨Examples⇨01.Basics ⇨Fade to call up the Fade sketch. The complete code for the Fade sketch follows:

```
/*
  Fade

  This example shows how to fade an LED on pin 9 using the analogWrite()
  function.

The analogWrite() function uses PWM, so if you want to change the pin you're
   using, be sure to use another PWM capable pin. On most Arduino, the PWM pins
   are identified with a "~" sign, like ~3, ~5, ~6, ~9, ~10 and ~11.

  This example code is in the public domain.

  http://www.arduino.cc/en/Tutorial/Fade
*/

int led = 9;            // the PWM pin the LED is attached to
int brightness = 0;     // how bright the LED is
int fadeAmount = 5;     // how many points to fade the LED by

// the setup routine runs once when you press reset:
void setup() {
  // declare pin 9 to be an output:
  pinMode(led, OUTPUT);
}

// the loop routine runs over and over again forever:
void loop() {
  // set the brightness of pin 9:
  analogWrite(led, brightness);
```

```
  // change the brightness for next time through the loop:
  brightness = brightness + fadeAmount;

  // reverse the direction of the fading at the ends of the fade:
  if (brightness <= 0 || brightness >= 255) {
    fadeAmount = -fadeAmount;
  }
  // wait for 30 milliseconds to see the dimming effect
  delay(30);
}
```

Upload this sketch to your board following the instructions at the start of the chapter. If everything has uploaded successfully, the LED fades from off to full brightness and then back off again.

If you don't see any fading, double-check your wiring:

- Make sure that you're using the correct pin number.
- Check that your LED is correctly situated, with the long leg connected by a wire to pin 9 and the short leg connected via the resistor and a wire to GND (ground).
- Check the connections on the breadboard. If the jump wires or components are not connected using the correct rows in the breadboard, they will not work.

Understanding the Fade sketch

By the light of your fading LED, take a look at how this sketch works.

The comments at the top of the sketch reveal exactly what's happening in this sketch: Using pin 9, a new function called `analogWrite()` causes the LED to fade off and on. After the comments, three declarations appear:

```
int led = 9;           // the PWM pin the LED is attached to
int brightness = 0;    // how bright the LED is
int fadeAmount = 5;    // how many points to fade the LED by
```

"But what is a declaration?" I hear you ask. Read on to find out.

Declarations

Declarations (which aren't something you put up at Christmas, ho ho ho) are values stored for later use by the program. In this case, three variables are declared, but you could declare many other variables or even include libraries of code in your sketch. For now, just remember that variables are declared before the `setup` function.

Variables

Variables are values that can change depending on what the program does with them. In C, you can declare the type, name, and value of the variable before the main body of code, much as ingredients are listed at the start of a recipe:

```
int led = 9;
```

The first part sets the type of the variable, creating an integer (`int`). An integer is a whole number, positive or negative, so no decimal places are required. Note that Arduino has lower and upper limits for the `int` variable: –32,768 to 32,767. Beyond those limits, a different type of variable must be used, known as a `long`. (You learn more about `long` variables in Chapter 10.) For now, an `int` will do just fine. The name of the variable is `led` and is purely for reference; it can be any single word that's useful for figuring out what the variable applies to. Finally, the value of the variable is set to 9, which is the number of the pin that is being used.

Variables are especially useful when you refer to a value repeatedly. In this case, the variable is called `led` because it refers to the pin to which the physical LED is attached. Now, every time you want to refer to pin 9, you can write `led` instead. Although this approach may seem like extra work initially, if you decided to change the pin to, say, pin 11, you would need to change only the variable at the start; every subsequent mention of `led` would automatically be updated. That's a big timesaver over having to trawl through the code to update every occurrence of 9.

The Fade sketch has three variables: `led`, `brightness`, and `fadeAmount`. These are integer variables and are capable of the same range of values, but each one is used for a different part of the process of fading an LED.

TIP

With declarations made, the code enters the `setup` function. The comments are reminders that `setup` runs only once and that just one pin is set as an output. Here you can see the first variable at work. Instead of writing `pinMode(9, OUTPUT)`, you have `pinMode(led, OUTPUT)`. Both work the same, but the latter uses the `led` variable:

```
// the setup routine runs once when you press reset:
void setup() {
  // declare pin 9 to be an output:
  pinMode(led, OUTPUT);
}
```

Then the loop starts to get a bit more complicated:

```
// the loop routine runs over and over again forever:
void loop() {
  // set the brightness of pin 9:
  analogWrite(led, brightness);

  // change the brightness for next time through the loop:
  brightness = brightness + fadeAmount;

  // reverse the direction of the fading at the ends of the fade:
  if (brightness <= 0 || brightness >= 255) {
    fadeAmount = -fadeAmount;
  }
  // wait for 30 milliseconds to see the dimming effect
  delay(30);
}
```

Instead of just on and off values, a fade needs a range of values. `analogWrite` allows you to send a value of 0 to 255 to a PWM pin on the Arduino. 0 is equal to 0v and 255 is equal to 5v, and any value in between gives a proportional voltage, thus fading the LED.

The loop begins by writing the `brightness` value to pin 9. A `brightness` value of 0 means that the LED is currently off:

```
  // set the brightness of pin 9:
  analogWrite(led, brightness);
```

Next you add the fade amount to the `brightness` variable, making it equal to 5. This value won't be written to pin 9 until the next loop:

```
// change the brightness for next time through the loop:
  brightness = brightness + fadeAmount;
```

The brightness must stay within the range that the LED can understand. This is accomplished by using an `if` statement that tests the value of the brightness variables to determine what to do next.

The word `if` starts the statement. The conditions are in the parentheses that follow, so in this case you have two. Is `brightness` less than or equal to 0 (indicated by the <= symbol) is the first condition. Is `brightness` greater than or equal to 255 (indicated by the >= symbol) is the second condition. In between the two conditional statements is the || symbol, which is the symbol for `OR`:

```
// reverse the direction of the fading at the ends of the fade:
  if (brightness <= 0 || brightness >= 255) {
    fadeAmount = -fadeAmount;
  }
```

So the complete statement is, "If the variable named brightness is less than or equal to 0, or greater than or equal to 255, do whatever is inside the curly brackets." When this eventually becomes true, the line of code inside the curly brackets is read. This basic mathematical statement inverts the `fadeAmount` variable. During the fade up to full brightness, 5 is added to the brightness with every loop. When 255 is reached, the `if` statement becomes true and `fadeAmount` changes from 5 to -5. Then every loop updates to "add minus 5" to the brightness until 0 is reached, when the `if` statement becomes true again. This inverts the `fadeAmount` of -5 to 5 to bring everything back to where it started:

```
    fadeAmount = -fadeAmount;
```

These conditions give us a number continually counting up and then down that the Arduino can use to continually fade your LED on and then off again.

Tweaking the Fade sketch

You can accomplish a task on your Arduino board in many ways. In this section, I will you one different way to fade an LED using the circuit that you created in the preceding section. The following code is the Fading example. From the Arduino menu, choose File⇨Examples⇨03.Analog⇨Fading to open the Fading sketch. Upload it and you will see that no visible difference exists between this and the preceding example.

REMEMBER

Some areas of the code appear colored on your screen, most often turquoise, green, or orange. This coloring marks a function or a statement recognized by the Arduino environment (and can be extremely handy for spotting typos). Color is impossible to recreate in a black-and-white book, so any colored code appears here in **bold**.

```
/*
  Fading

  This example shows how to fade an LED using the analogWrite() function.

  The circuit:
  - LED attached from digital pin 9 to ground.

  created 1 Nov 2008
  by David A. Mellis
  modified 30 Aug 2011
  by Tom Igoe

  This example code is in the public domain.

  http://www.arduino.cc/en/Tutorial/Fading
*/

int ledPin = 9;  // LED connected to digital pin 9

void setup() {
  // nothing happens in setup
}

void loop() {
  // fade in from min to max in increments of 5 points:
  for(int fadeValue = 0;fadeValue <= 255;fadeValue += 5) {
    // sets the value (range from 0 to 255):
    analogWrite(ledPin, fadeValue);
    // wait for 30 milliseconds to see the dimming effect
    delay(30);
  }

  // fade out from max to min in increments of 5 points:
  for(int fadeValue = 255 ; fadeValue >= 0; fadeValue -= 5) {
   // sets the value (range from 0 to 255):
   analogWrite(ledPin, fadeValue);
   // wait for 30 milliseconds to see the dimming effect
   delay(30);
  }
}
```

The Fade example is efficient and does a simple fade very well, but it relies on the `loop` function to update the LED value. This version uses `for` loops, which operate within the main Arduino `loop` function.

Using for loops

When a sketch enters a `for` loop, it sets up the criteria for exiting the loop and can't move out of it until the criteria are met. `for` loops are often used for repetitive operations; in this case, `for` loops are used to increase or decrease a number at a set rate to create the repeating fade.

The first line of the `for` loop defines the initialization, the test, and the amount of increment or decrement:

```
for(int fadeValue = 0;fadeValue <= 255;fadeValue += 5)
```

In plain English, this reads: "Make a variable called `fadeValue` (that is local to this `for` loop) equal to a value of 0; check to see whether it is less than or equal to 255; if it is, set `fadeValue` to be equal to `fadeValue` plus 5." `fadeValue` is equal to 0 only when it is created; after that, it is increased by 5 every time the `for` loop cycles.

Within the loop, the code updates the `analogWrite` value of the LED and waits 30 milliseconds (ms) before attempting the loop one more time:

```
for (int fadeValue = 0 ; fadeValue <= 255; fadeValue += 5) {
  // sets the value (range from 0 to 255):
  analogWrite(ledPin, fadeValue);
  // wait for 30 milliseconds to see the dimming effect
  delay(30);
}
```

This `for` loop behaves in the same way as the main `loop` in the default Fade example, but because the `fadeValue` is contained in its own loop, and broken into fade up and fade down loops, it is a lot easier to experiment with fading patterns in a more controlled way. For example, change `+= 5` and `-= 5` to different values (that divide into 255 neatly) for some interesting asymmetrical fading.

You could also copy and paste the same `for` loops to create further fading animations. However, bear in mind that your Arduino can do nothing else while it is in a `for` loop.

The Button Sketch

The first and perhaps most basic of inputs that you can and should learn for your Arduino projects is the modest pushbutton.

For the Button project, you need the following:

- An Arduino Uno
- A breadboard
- A 10k ohm resistor
- A pushbutton
- An LED
- Jump wires

Figure 6-3 shows the breadboard layout for the Button circuit. It's important to note which legs of the pushbutton are connected. In most cases, these small pushbuttons are made to bridge the gap over the center of your breadboard exactly. If they do bridge the gap, the legs are usually split at 90 degrees to the gap (left to right on this diagram).

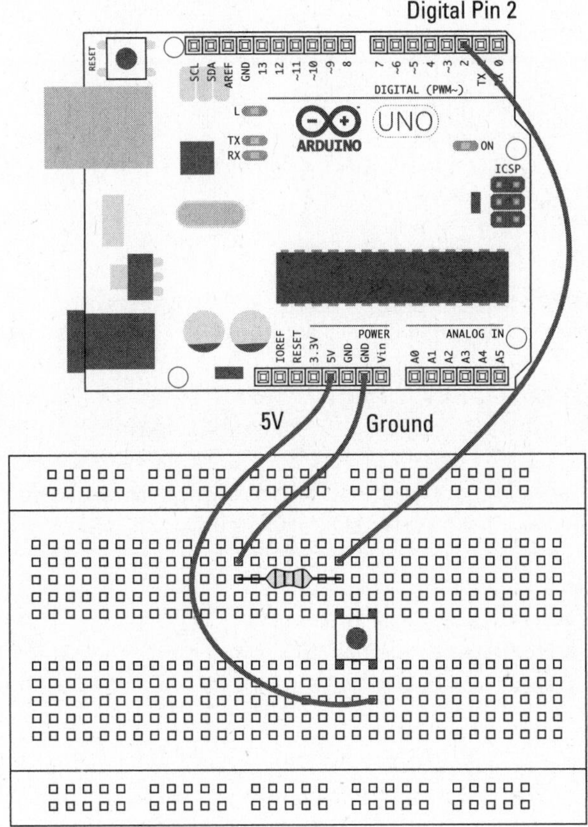

FIGURE 6-3: Pin 2 is reading the pushbutton.

You can test the legs of a pushbutton with a continuity tester if your multimeter has that function (as detailed in Chapter 4).

From the schematic in Figure 6-4, you can see that the resistor leading to ground should be connected to the same side of the pushbutton as pin 2, and that when the pushbutton is pressed, it connects the 5V pin to both D2 and GND. This setup is used to compare ground (0V) to a voltage (5V) so that you can tell whether the switch is open or closed.

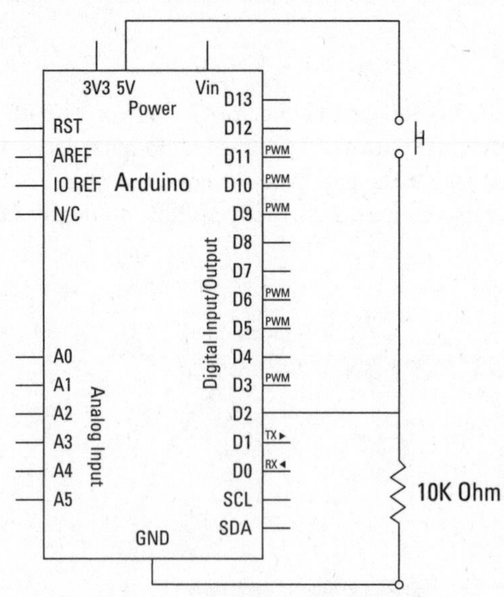

FIGURE 6-4: A schematic of the pushbutton circuit.

Build the circuit and upload the code from File ➪ Examples ➪ 02.Digital ➪ Button.

```
/*
  Button

  Turns on and off a light emitting diode(LED) connected to digital pin 13,
  when pressing a pushbutton attached to pin 2.

  The circuit:
  - LED attached from pin 13 to ground
  - pushbutton attached to pin 2 from +5V
  - 10K resistor attached to pin 2 from ground
```

```
  - Note: on most Arduinos there is already an LED on the board
    attached to pin 13.

  created 2005
  by DojoDave <http://www.0j0.org>
  modified 30 Aug 2011
  by Tom Igoe

  This example code is in the public domain.

  http://www.arduino.cc/en/Tutorial/Button
*/

// constants won't change. They're used here to
// set pin numbers:
const int buttonPin = 2;    // the number of the pushbutton pin
const int ledPin = 13;    // the number of the LED pin

// variables will change:
int buttonState = 0;       // variable for reading the pushbutton status

void setup() {
 // initialize the LED pin as an output:
 pinMode(ledPin, OUTPUT);
 // initialize the pushbutton pin as an input:
 pinMode(buttonPin, INPUT);
}

void loop(){
 // read the state of the pushbutton value:
 buttonState = digitalRead(buttonPin);

 // check if the pushbutton is pressed.
 // if it is, the buttonState is HIGH:
 if (buttonState == HIGH) {
  // turn LED on:
  digitalWrite(ledPin, HIGH);
 }
 else {
  // turn LED off:
  digitalWrite(ledPin, LOW);
 }
}
```

After you upload the sketch, give your button a press and you should see the pin 13 LED light. (To make the light easier to see, add a bigger LED to your Arduino board between pin 13 and GND.)

If you don't see anything lighting up, double-check your wiring:

- Make sure that your button is connected to the correct pin number.
- If you're using an additional LED, check that it is correctly situated, with the long leg in pin 13 and the short leg in GND. You can also remove it and monitor the LED mounted on the board (marked L) instead.
- Check the connections on the breadboard. Make sure that the jump wires and components are connected using the correct rows in the breadboard.

Understanding the Button sketch

This sketch is your first interactive Arduino project. The previous sketches were all about outputs, but now you can affect those outputs by providing real-world human input!

While pressed, your button turns on a light. When the button is released, the light turns off. Let's take a look at the sketch from the top to see how this happens.

The first step in the Button sketch is to declare constants and variables. A *constant*, displayed as `const`, is a value or an identifier whose value doesn't change for the duration of the program. In the next example the values of `buttonPin` and `ledPin` are both defined as constants. This approach is best used for values that aren't supposed to change; this way, you are making doubly sure that they won't. Pin numbers are being assigned because you won't change the pin number physically.

The variable `buttonState` is set to 0. This variable monitors changes to the button:

```
const int buttonPin = 2;     // the number of the pushbutton pin
const int ledPin = 13;     // the number of the LED pin

// variables will change:
int buttonState = 0;       // variable for reading the pushbutton status
```

`setup` establishes `pinMode`, with `ledPin` (pin 13) as the output and `buttonPin`,(pin 2) as the input:

```
void setup() {
  // initialize the LED pin as an output:
  pinMode(ledPin, OUTPUT);
  // initialize the pushbutton pin as an input:
  pinMode(buttonPin, INPUT);
}
```

In the main loop, you can see the order of things clearly. First, the `digitalRead` function is used on pin 2. Just as `digitalWrite` can write a `HIGH` or `LOW` (1 or 0) value to a pin, `digitalRead` can read the current value, either HIGH (1) or LOW (0), from a pin. That value is then stored in the `buttonState` variable:

```
void loop(){
  // read the state of the pushbutton value:
  buttonState = digitalRead(buttonPin);
```

With the button state established, a test uses an `if` statement to determine what happens next. The statement reads: "If there is a `HIGH` value (voltage connected to the circuit), then send a `HIGH` value to `ledPin` (pin 13) to turn the LED on; if there is a `LOW` value (the pin is grounded), then send a `LOW` value to `ledPin` to turn the LED off; repeat."

```
  // check if the pushbutton is pressed.
  // if it is, the buttonState is HIGH:
  if (buttonState == HIGH) {
    // turn LED on:
    digitalWrite(ledPin, HIGH);
  }
  else {
    // turn LED off:
    digitalWrite(ledPin, LOW);
  }
}
```

Tweaking the Button sketch

It's often necessary to invert the output of a switch or sensor, and you can do this in two ways. The easiest is to change one word in the code.

By changing the line of code in the preceding sketch from

```
  if (buttonState == HIGH)
```

to

```
if (buttonState == LOW)
```

the output is reversed. This means that the LED is on until the button is pressed. If you have a computer, this option is the easiest. Simply upload the code.

However, sometimes (such as when your laptop battery is dead) you don't have the means to upload the edited code. Often, the easiest way to flip the logic is to flip the polarity of the circuit.

Instead of connecting pin 2 to a resistor and then GND, connect that resistor to 5V and move the GND wire to the other side of the button, as shown in Figure 6-5.

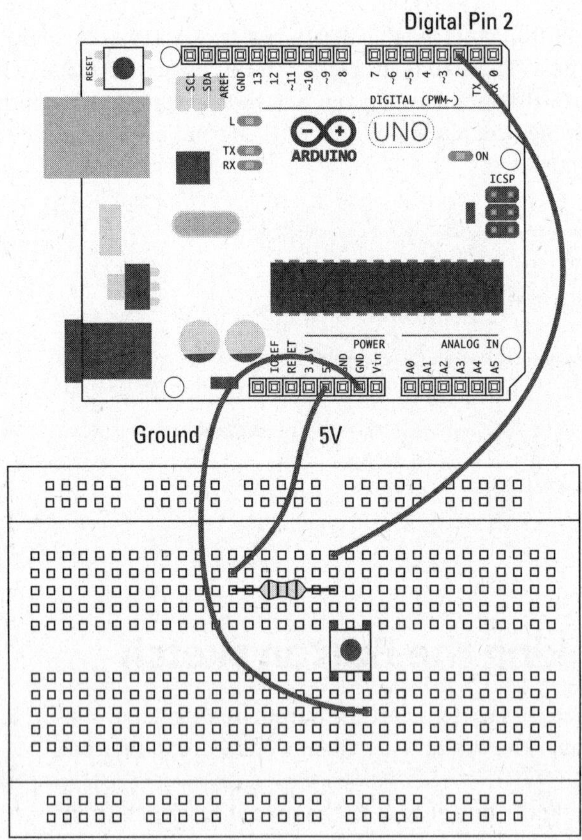

FIGURE 6-5: A button with the polarity flipped.

The AnalogInput Sketch

The preceding sketch showed you how to use `digitalRead` to read either on or off, but what if you want to handle an analog value such as a dimmer switch or a volume control knob?

For the AnalogInput project, you need

- An Arduino Uno
- A breadboard
- A 10k ohm variable resistor
- An LED
- Jump wires

VARIABLE RESISTORS

Variable resistors (also known as *potentiometers* or *pots*), like standard passive resistors, resist the flow of current in a circuit. The difference is that rather than having a fixed value, they have a range. Normally, the upper limit of this range is printed on the resistor. For example, a variable resistor with a value of 10K Ω gives you a range of 0 ohms to 10,000 ohms. This change is something that can be monitored electrically to give a variable analog input.

Variable resistors come in a variety of shapes and sizes, as shown in the following figure. Anything with this analog movement, such as a thermostat, the dial on your washing machine, or the dial on your toaster to set the time, most likely contains a potentiometer.

In Figure 6-6 you see the layout for this circuit. You need an LED and a resistor for your output, and a variable resistor for your input. If you're not quite sure what a variable resistor is, check out the "Variable Resistors" sidebar.

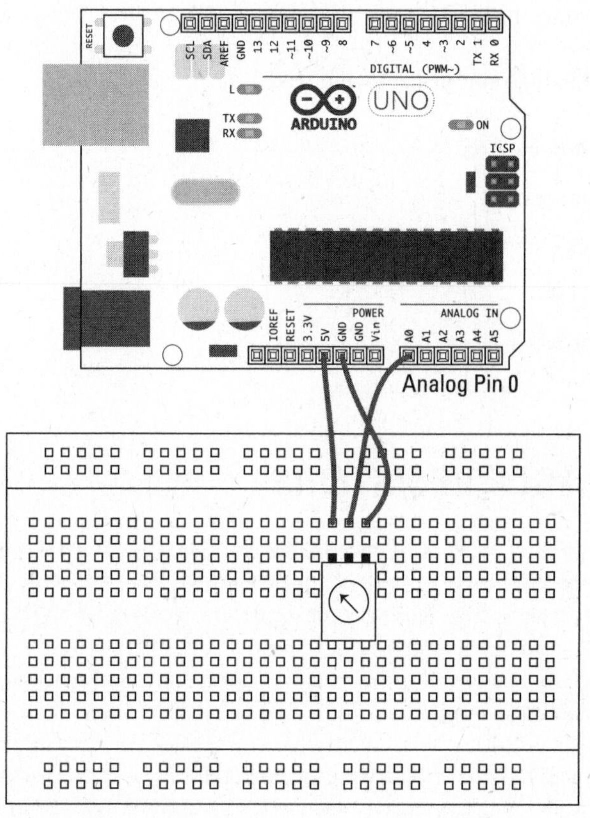

FIGURE 6-6: The potentiometer is connected to Analog Pin 0.

In Figures 6-6 and 6-7, the variable resistor has power and ground connected across opposite pins, with the central pin providing the reading. To read the analog input, you need to use the special set of analog input pins on the Arduino board.

TIP

Note that if you were to swap the resistor's polarity (the positive and negative wires), you would invert the direction of the potentiometer. This can be a quick fix if you find that your resistor's value is going in the wrong direction.

Build the circuit in Figure 6-6, and then upload the code from File ⇨ Examples ⇨ 03. Analog ⇨ AnalogInput to start controlling your variable resistor.

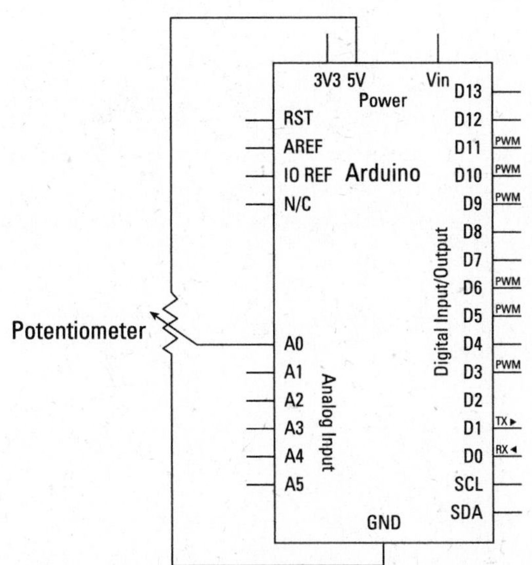

FIGURE 6-7: A schematic of the AnalogInput circuit.

```
/*
  Analog Input

  Demonstrates analog input by reading an analog sensor on analog pin 0 and
  turning on and off a light emitting diode(LED) connected to digital pin 13.
  The amount of time the LED will be on and off depends on the value obtained
  by analogRead().

  The circuit:
  - potentiometer
    center pin of the potentiometer to the analog input 0
    one side pin (either one) to ground
    the other side pin to +5V
  - LED
    anode (long leg) attached to digital output 13
    cathode (short leg) attached to ground

  - Note: because most Arduinos have a built-in LED attached to pin 13 on the
    board, the LED is optional.

  created by David Cuartielles
  modified 30 Aug 2011
  By Tom Igoe

  This example code is in the public domain.
```

```
  http://www.arduino.cc/en/Tutorial/AnalogInput
*/

int sensorPin = A0;  // select the input pin for the potentiometer
int ledPin = 13;    // select the pin for the LED
int sensorValue = 0; // variable to store the value coming from the sensor

void setup() {
 // declare the ledPin as an OUTPUT:
 pinMode(ledPin, OUTPUT);
}

void loop() {
 // read the value from the sensor:
 sensorValue = analogRead(sensorPin);
 // turn the ledPin on
 digitalWrite(ledPin, HIGH);
 // stop the program for <sensorValue> milliseconds:
 delay(sensorValue);
 // turn the ledPin off:
 digitalWrite(ledPin, LOW);
 // stop the program for <sensorValue> milliseconds:
 delay(sensorValue);
}
```

After the sketch is uploaded, turn the potentiometer. The result is an LED that blinks slower or faster depending on the value of the potentiometer. You can add another LED between pin 13 and GND to improve the effect of this spectacle.

If you don't see anything lighting up, double-check your wiring:

- » Make sure that you're using the correct pin number for your variable resistor.
- » Check that your LED is the correct way around, with the long leg in pin 13 and the short leg in GND.
- » Check the connections on the breadboard. If the jump wires or components are not connected using the correct rows in the breadboard, they will not work.

Understanding the AnalogInput sketch

Analog sensors come in a variety of forms, but the principle is generally the same for each of them. In this section you examine the AnalogInput sketch to get a better understanding of how Arduino interprets these sensors.

The declarations state the pins that this sketch uses. The pin for the analog reading is written as `A0`, which is short for analog input pin 0 to mark it as the first analog input pin in the row of 6 (numbered 0 to 5). Both `ledPin` and `sensorValue` are declared as standard variables. It's worth noting that `ledPin` and `sensorPin` could both be declared as constant integers (`const`) because they don't change. Because the `sensorValue` value will change, it is stored as a variable:

```
int sensorPin = A0;   // select the input pin for the potentiometer
int ledPin = 13;     // select the pin for the LED
int sensorValue = 0; // variable to store the value coming from the sensor
```

During setup, you need only declare the `pinMode` of the digital `ledPin`. The analog input pins, as their name implies, are for input only.

TIP

You can use the analog input pins also as more basic digital input or output pins. Instead of referring to them as analog pins A0–A5, you could number them as digital pins 14–19, as an extension of the existing digital pins. Each must then be declared as either an input or a n output using the `pinMode` function, as with any digital pin:

```
void setup() {
 // declare the ledPin as an OUTPUT:
 pinMode(ledPin, OUTPUT);
}
```

Similarly to the Button sketch, the AnalogInput sketch reads the sensor first. When using the `analogRead` function, it interprets the voltage value of an analog pin. As the resistance changes, so too does the voltage. The accuracy of the value depends on the quality of the variable resistor. The Arduino uses the analog-to-digital converter on the ATmega328 chip to read this analog voltage. Instead of 0V, 0.1V, 0.2V, and so on, the Arduino returns a value as an integer in the range of 0–1023. For example, a voltage of 2.5V is interpreted as 511.

It's generally a good idea to read the sensor data first to prevent any delays when reading values, even though the looping occurs extremely quickly. Otherwise, this can give the effect of a lag in the response of the sensor.

After `sensorValue` is read, the sketch is essentially the same as the Blink sketch, but with a variable delay. The `ledPin` is written `HIGH`, waits, is written `LOW`, waits for the same amount of time, updates the sensor value, and repeats.

Using the raw sensor value (0–1023) makes the delay between 0 seconds and 1.023 seconds:

```
void loop() {
  // read the value from the sensor:
  sensorValue = analogRead(sensorPin);
  // turn the ledPin on
  digitalWrite(ledPin, HIGH);
  // stop the program for <sensorValue> milliseconds:
  delay(sensorValue);
  // turn the ledPin off:
  digitalWrite(ledPin, LOW);
  // stop the program for <sensorValue> milliseconds:
  delay(sensorValue);
}
```

This sketch blinks your LED at various rates. However, as the blinks become slower, the delays in the loop become longer and, therefore, the readings from the sensor become less frequent. This can make your sensor less responsive when it's at the higher values, giving you less consistent readings. For another look at sensors, as well as how to smooth and calibrate them, head over to Chapter 10.

Tweaking the AnalogInput sketch

The analogRead function has supplied an integer value, and you can apply all sorts of conditions or calculations to that number in your sketch. In this example, I show you how to measure whether a sensor value is above a certain number or threshold.

By putting an if statement around the digitalWrite part of the loop, you are able to set a threshold. In this example, the LED blinks only if it is over the sensor's halfway value of 511:

```
void loop() {
  // read the value from the sensor:
  sensorValue = analogRead(sensorPin);
 if (sensorValue > 511){
  // turn the ledPin on
  digitalWrite(ledPin, HIGH);
  // stop the program for <sensorValue> milliseconds:
  delay(sensorValue);
  // turn the ledPin off:
```

```
    digitalWrite(ledPin, LOW);
    // stop the program for <sensorValue> milliseconds:
    delay(sensorValue);
  }
}
```

Try adding some conditions, but be aware that the sensor won't update as frequently if there are too many delays. For other sketches that remedy this problem, check out the BlinkWithoutDelay sketch in Chapter 10.

Talking Serial

It's good to see the effects of your circuit through an LED, but unless you can see the values, you might find it difficult to know whether a circuit is behaving as expected. The project in this section and the one following are designed to display the value of inputs by using the *serial monitor*

Serial is a method of communication between a peripheral and a computer. In this case, it is serial communication over Universal Serial Bus (USB). Data is sent one byte at a time in the order that it is written. When reading sensors with an Arduino, the values are sent over this connection and can be monitored or interpreted on your computer.

The DigitalReadSerial Sketch

In the DigitalReadSerial project, you monitor the `HIGH` and `LOW` values of a button over the serial monitor.

For this project, you need the following:

- An Arduino Uno
- A breadboard
- A 10k ohm resistor
- A pushbutton
- Jump wires

Figures 6-8 and 6-9 use the same circuit as you use for the Button sketch described previously in this chapter, but note that there are some slight changes to the code for this project.

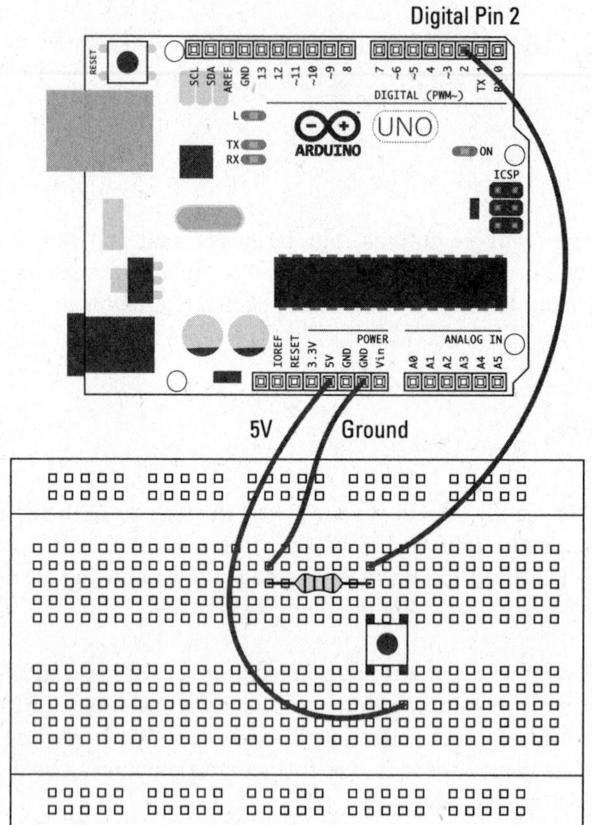

FIGURE 6-8: Pin 2 is reading the pushbutton.

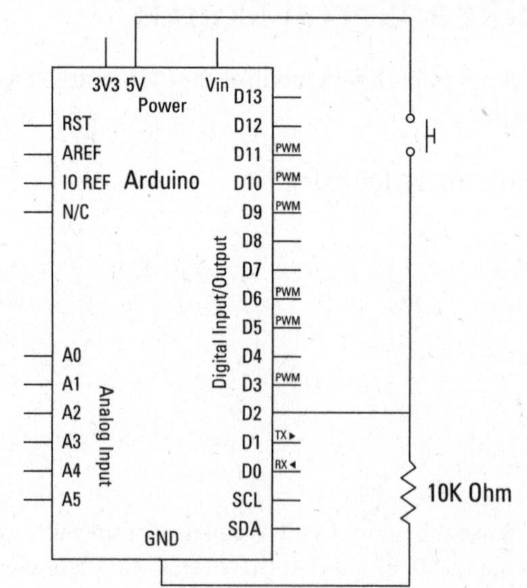

FIGURE 6-9: A schematic of the pushbutton circuit.

Complete the circuit and upload the code from File ⇨ Examples ⇨ 01. Basics ⇨ DigitalReadSerial:

```
/*
  DigitalReadSerial

  Reads a digital input on pin 2, prints the result to the Serial Monitor

  This example code is in the public domain.

  http://www.arduino.cc/en/Tutorial/DigitalReadSerial
*/

// digital pin 2 has a pushbutton attached to it. Give it a name:
int pushButton = 2;

// the setup routine runs once when you press reset:
void setup() {
  // initialize serial communication at 9600 bits per second:
  Serial.begin(9600);
  // make the pushbutton's pin an input:
  pinMode(pushButton, INPUT);
}

// the loop routine runs over and over again forever:
void loop() {
  // read the input pin:
  int buttonState = digitalRead(pushButton);
  // print out the state of the button:
  Serial.println(buttonState);
  delay(1);      // delay in between reads for stability
}
```

After you upload the sketch, click the serial monitor button on the top right of the Arduino window. Clicking this button opens the serial monitor window shown in Figure 6-10 and displays any values being sent to the currently selected serial port (which is the same as the one you have just uploaded to unless you selected otherwise).

You should see a cascade of `0` values in the window. Press the button a few times, and some `1` values should appear.

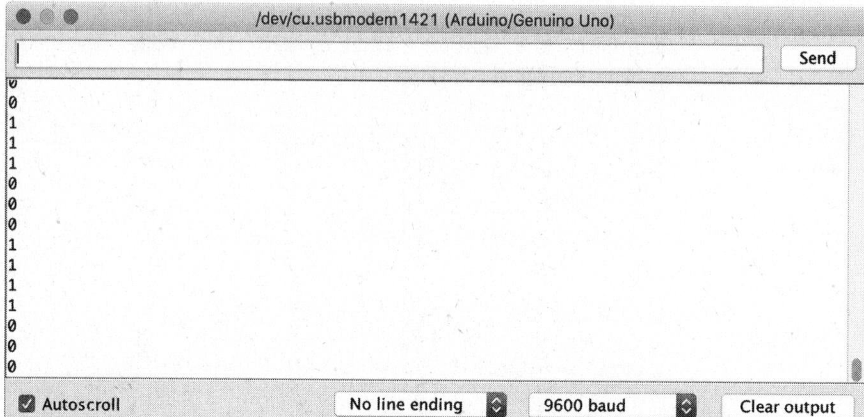

FIGURE 6-10: The serial monitor window is great for keeping an eye on whatever your Arduino is up to.

If you don't see anything, or you see incorrect values, double-check your wiring:

- » Make sure that you're using the correct pin number for your button.
- » Check the connections on the breadboard. If the jump wires or components are not connected using the correct rows in the breadboard, they will not work.
- » If you receive strange characters instead of 0s and 1s, check the baud rate in the serial monitor; if it is not set to 9600, use the drop-down menu to select that rate.

Understanding the DigitalReadSerial sketch

The only variable to declare in this sketch is the pin number for the pushbutton:

```
int pushButton = 2;
```

In the `setup` function, there is a new function called `Serial.begin`. This function initializes serial communication. The number in parentheses represents the speed of the communication, or *baud rate,* and is the number of bits sent per second. In this case, the Arduino is sending data at 9600 bits per second.

When viewing the communication in the serial monitor, it is important to read the data at the same rate it is being written. If you don't, the data is scrambled and you are presented with what looks like gibberish. In the bottom right of the window, you can set the baud rate, but by default it should be set to 9600.

```
void setup() {
  // initialize serial communication at 9600 bits per second:
  Serial.begin(9600);
  // make the pushbutton's pin an input:
  pinMode(pushButton, INPUT);
}
```

In the loop, pushButton is read and its value stored in the buttonState variable:

```
void loop() {
  // read the input pin:
  int buttonState = digitalRead(pushButton);
```

The Serial.println function then writes the value of to the serial port. When println is used, it signifies that a carriage return (new line) should be added after the value is printed. The carriage return is especially useful when you're reading values because they appear much clearer than they do in one line of values:

```
  // print out the state of the button:
  Serial.println(buttonState);
```

A delay of 1ms is added to the end of the loop to slow the speed at which the button is read. Having values written faster than they can be displayed can cause unstable results, so it's advisable to keep this delay:

```
  delay(1);      // delay in between reads for stability
}
```

The AnalogInOutSerial Sketch

In the AnalogInOutSerial project, you monitor an analog value sent by a variable resistor over the serial monitor. These variable resistors are the same as the volume control knobs on your stereo, but people often have no idea how they work. In this example, you monitor the value as detected by your Arduino and display it on your screen in the serial monitor, giving you a greater understanding of the range of values and the performance of this analog sensor.

You need the following:

- An Arduino Uno
- A breadboard
- A 10k ohm variable resistor

» A resistor (greater than 120 ohm)

» An LED

» Jump wires

The circuit, as shown in Figures 6-11 and 6-12, is similar to the example for the `AnalogInput` circuit, but with the addition of an LED connected to pin 9 as in the Fade circuit. The code fades the LED on and off according to the turn of the potentiometer. Because the input and the output have a different range of values, the sketch includes a conversion to use the potentiometer to fade the LED. This circuit is a great example of using the serial monitor for debugging and displays both the input and output values for maximum clarity.

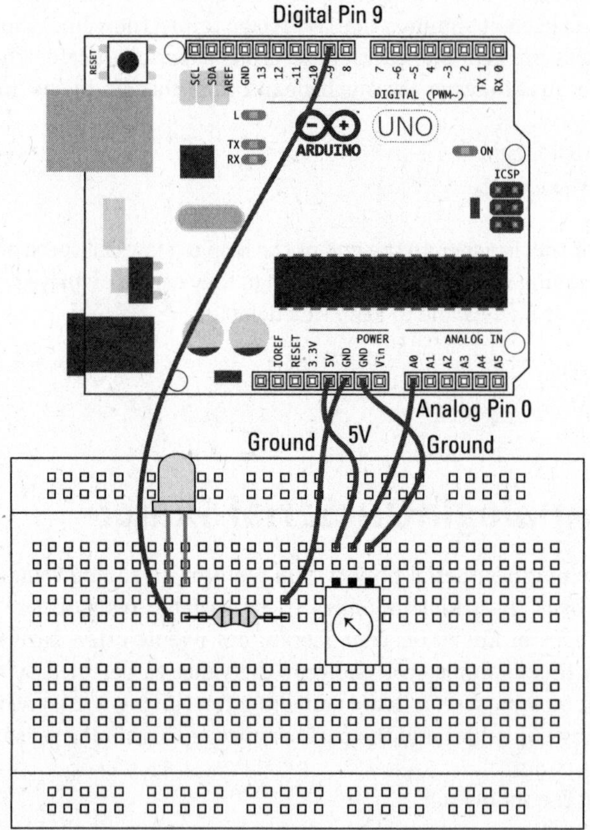

FIGURE 6-11: Dimmer circuit to be read over serial.

Complete the circuit and upload the code from File⇨Examples⇨03. Analog⇨AnalogInOutSerial.

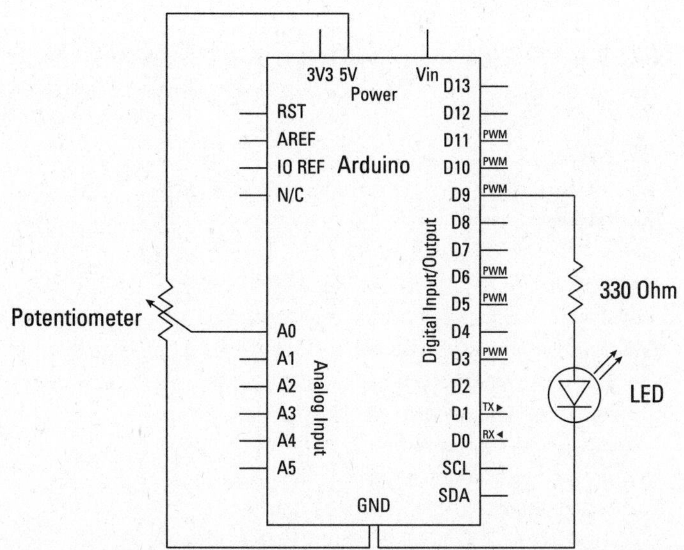

FIGURE 6-12: A schematic of the dimmer circuit.

```
/*
  Analog input, analog output, serial output

  Reads an analog input pin, maps the result to a range from 0 to 255 and uses
  the result to set the pulse width modulation (PWM) of an output pin.
  Also prints the results to the Serial Monitor.

  The circuit:
  - potentiometer connected to analog pin 0.
    Center pin of the potentiometer goes to the analog pin.
    side pins of the potentiometer go to +5V and ground
  - LED connected from digital pin 9 to ground

  created 29 Dec. 2008
  modified 9 Apr 2012
  by Tom Igoe

  This example code is in the public domain.

  http://www.arduino.cc/en/Tutorial/AnalogInOutSerial
*/

// These constants won't change. They're used to give names
// to the pins used:
const int analogInPin = A0;  // Analog input pin that the potentiometer is
   attached to
const int analogOutPin = 9; // Analog output pin that the LED is attached to
```

```
int sensorValue = 0;        // value read from the pot
int outputValue = 0;        // value output to the PWM (analog out)

void setup() {
  // initialize serial communications at 9600 bps:
  Serial.begin(9600);
}

void loop() {
  // read the analog in value:
  sensorValue = analogRead(analogInPin);
  // map it to the range of the analog out:
  outputValue = map(sensorValue, 0, 1023, 0, 255);
  // change the analog out value:
  analogWrite(analogOutPin, outputValue);

  // print the results to the serial monitor:
  Serial.print("sensor = " );
  Serial.print(sensorValue);
  Serial.print("\t output = ");
  Serial.println(outputValue);

  // wait 2 milliseconds before the next loop
  // for the analog-to-digital converter to settle
  // after the last reading:
  delay(2);
}
```

After you upload the sketch, turn the potentiometer with your fingers. The result should be an LED that fades on and off depending on the value of the potentiometer. Now click the serial monitor button on the top right of the Arduino window to monitor the numerical values you are receiving and sending to the LED.

If you don't see anything happening, double-check your wiring:

- » Make sure that you're using the correct pin number for your variable resistor.
- » Check that your LED is the correct way round, with the long leg connected to Pin 9 and the short leg connected to GND, via a resistor.
- » Check the connections on the breadboard. The jump wires and components will not work if you are not using the correct rows in the breadboard.
- » If you receive strange characters instead of words and numbers, check the baud rate in the serial monitor. If it is not set to 9600, use the drop-down menu to select that rate.

Understanding the AnalogInOutSerial sketch

The start of the sketch is straightforward. It declares constants for both of the pins in use for the analog input and the PWM output. There are also two variables for the raw data from the sensor (sensorValue) and the value that is sent to the LED (outputValue):

```
const int analogInPin = A0; // Analog input pin that the potentiometer is
            attached to
const int analogOutPin = 9; // Analog output pin that the LED is attached to

int sensorValue = 0;     // value read from the pot
int outputValue = 0;     // value output to the PWM (analog out)
```

In the setup function, you have little to do beyond opening the serial communication line:

```
void setup() {
 // initialize serial communications at 9600 bps:
 Serial.begin(9600);
}
```

The real action is in the loop. As in the Fade sketch, the best place to start is with reading the input. The sensorValue variable stores the reading from analog InPin, which will be in the range of 0–1024:

```
void loop() {
 // read the analog in value:
 sensorValue = analogRead(analogInPin);
```

Because fading an LED using PWM requires a range of 0–255, you need to scale the sensorValue down to make it fit the lower range of outputValue. To do so, you use the map function. The map function scales a variable. By setting the variable's minimum and maximum to a new minimum and maximum, all the scaling is handled for you. The map function creates an outputValue directly proportional to the sensorValue but on a smaller scale:

```
 // map it to the range of the analog out:
 outputValue = map(sensorValue, 0, 1023, 0, 255);
```

TIP

Functions like this are useful but can sometimes be overkill. In this example, you can achieve the same result by simply dividing the sensor value by 4:

```
outputValue = sensorValue/4;
```

Because outputValue is an integer, it is rounded to the nearest whole number. The outputValue is then written to the LED using the analogWrite function:

```
// change the analog out value:
analogWrite(analogOutPin, outputValue);
```

This code is enough for the circuit to function, but if you want to know what's going on, you need to write some values to the serial port. The code has three lines of Serial.print before Serial.println. Each occurrence of Serial.print writes text on the same line of the serial monitor. Using Serial.println finishes that text with a carriage return, starting a new line. Therefore, a new line of text is written every time the program completes a loop.

Text inside the quotation marks is for labeling or adding characters. You can also use special characters, such as \t, which adds a tab for spacing:

```
// print the results to the serial monitor:
Serial.print("sensor = " );
Serial.print(sensorValue);
Serial.print("\t output = ");
Serial.println(outputValue);
```

An example of this line in the serial monitor follows:

```
sensor = 1023    output = 511
```

The loop is finished with a short delay to stabilize the results and then the loop is repeated, updating the input, output, and readings on the serial monitor.

```
// wait 2 milliseconds before the next loop
// for the analog-to-digital converter to settle
// after the last reading:
delay(2);
}
```

TIP

This delay time is largely arbitrary. The 2 ms could just as well be 1 ms, as in the previous example. You may have to experiment with these small delays. If a sensor is jumpy, you may want to go up to 10 ms, or you may find that the reading is perfectly smooth and can be removed completely. There is no magic value.

» Switching bigger loads with transistors

» Speeding up your motor

» Turning with precision using a stepper motor

» Making electronic music with a buzzer

Chapter 7

More Basic Sketches: Motion and Sound

Chapter 6 shows you how to use some simple LEDs as outputs for various circuits. In Arduino land, nothing is more beautiful than a blinking LED, but you have a variety of other outputs as options. In this chapter, I explore two other areas: motion provided by motors and sound from a buzzer.

Working with Electric Motors

Electric motors allow you to move things with electricity using the power of *electromagnetism*. When an electrical current is passed through a coil of wire, it creates an electromagnetic field. This process works similarly to a normal permanent bar magnet but gives you control over the presence of the field, meaning that you can turn it on and off at will and even change the direction of the magnetism. As you may remember from school, magnets have two possible states: attraction or repulsion. In an electromagnetic field, you can switch between these by changing the polarity, which in practical terms means switching the positive and negative wires.

Electromagnets have a variety of uses, such as in electrically operated locks, automated plumbing valves, and read-write heads on hard disks. They're used also for lifting scrap metal. Even the CERN Large Hadron Collider uses an electromagnet. In this chapter, I focus on another important use: electric motors.

An electric motor is made up of a coil of wire (electromagnet) between two regular, permanent magnets. By alternating the polarity of the coil, it is possible to rotate it because it is pulled by one magnet and then pushed toward the next. If this is done fast enough, the coil gathers momentum to spin.

The first part to understand is how the coil can spin if it is attached to wires. This spin is achieved by mounting two copper brushes on the axle. The brushes stay in contact with two semicircles of copper, as shown in Figure 7-1, so a connection can be maintained without any fixed wires. The semicircles also mean that the two points are never in contact, which would cause a short circuit.

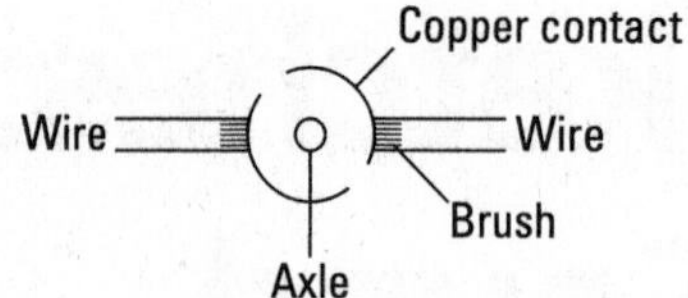

FIGURE 7-1: How a motor's axle can be connected but still be free to spin.

With a freely spinning coil in place on an axle, you can affect the coil by placing two permanent bar magnets near it. As shown in Figure 7-2, the magnets are placed on either side of the coil, with different poles on each side. If you put electrical current across the coil, you give it a polarity — either north or south, as with conventional bar magnets. If the coil is north, it is repelled by the north bar magnet and attracted by the south bar magnet.

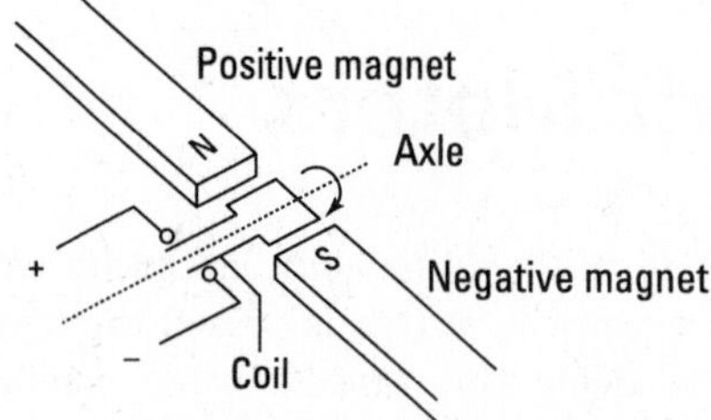

FIGURE 7-2: A diagram of an electric motor.

If you look at the brush again, you realize that something else happens when the coil does a half rotation: The polarity flips. When this happens, the cycle starts again and the north coil becomes south and is pushed away by the south magnet back to north. Because of the momentum produced when the coil is repelled, this movement continues in the same direction while sufficient power exists.

This type of electric motor is the most basic; modern ones are highly refined, with more coils and magnets to produce a smoother movement. Other motors are also based on this principle but have more advanced controls to move, for example, by a precise number of degrees or to a specific location. In your kit, you should have two varieties of electric motor: a DC motor and a servo motor.

Discovering Diodes

An essential component for motor control circuits is the diode. As explained earlier in this chapter, you can spin an electric motor by putting voltage through it. But if a motor is spinning or is turned without having a voltage put through it, it generates a voltage in the opposite direction; this is how electric generators and dynamos produce electricity from movement.

If this reversal of voltage happens in your circuit, the effects can be disastrous, including damaged or destroyed components. So to control this reverse current, you use a diode. *Diodes* block current in one direction and allow it in the other. Current can flow from the anode to the cathode. Figure 7-3 shows how the anode and cathode are marked for both the physical diode and the circuit diagram, with a band on the physical diode and a solid line on the schematic, both indicating the cathode.

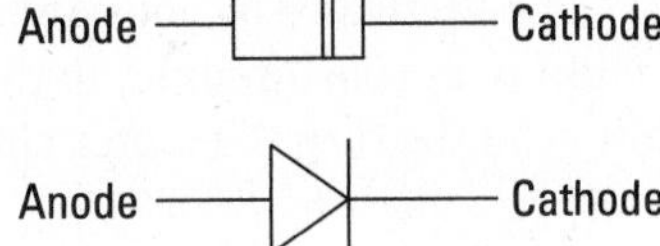

FIGURE 7-3: A physical diode and its schematic symbol.

Spinning a DC Motor

The DC motor in your kit (also known as a hobby motor or a brushed DC motor) is the most basic of electric motors and is used in all types of hobby electronics such as model planes and trains. When current is passed through a DC motor, it spins continuously in one direction until the current stops. Unless specifically marked with a + or –, DC motors have no polarity, which means that you can swap the two wires to reverse the direction of the motor. Many other, bigger motors exist, but in this example I stick to the small hobby motors.

The Motor sketch

In this section, I show you how to set up a simple control circuit to turn your motor on and off.

You need the following:

- An Arduino Uno
- A breadboard
- A transistor
- A DC motor
- A diode
- A 2.2k ohm resistor
- Jump wires

Figure 7-4 shows the layout for this circuit.

The circuit diagram in Figure 7-5 should clarify exactly what is going on. To power the motor, you need to send 5V through it and then on to ground. This voltage spins the motor, but you have control of it. To give your Arduino control of the motor's power, and therefore its rotation, you place a transistor just after the motor. The transistor, as described in the sidebar "Understanding Transistors," is an electrically operated switch that can be activated by your Arduino's digital pins. In this example, it is controlled by pin 9 on your Arduino, in the same way as an LED except that the transistor allows you the turn the motor circuit on and off.

This circuit works, but it still allows the chance of creating a reverse current because of the momentum of the motor as it slows down or because the motor could be turned. If reverse current is generated, it travels from the negative side of the motor and tries to find the easiest route to ground. This route may be through the transistor or through the Arduino. You can't know for sure what will happen, so you need to provide a way to control this excess current.

To be safe, you place a diode across the motor. The diode faces toward the source of the voltage, so the voltage is forced through the motor, which is what you want. If current is generated in the opposite direction, it will now be blocked from flowing into the Arduino.

WARNING

If you place the diode the wrong way, the current bypasses the motor and you create a short circuit. The short circuit tries to ground all the available current and could break your USB port or, at the very least, display a warning message, informing you that your USB port is drawing too much power.

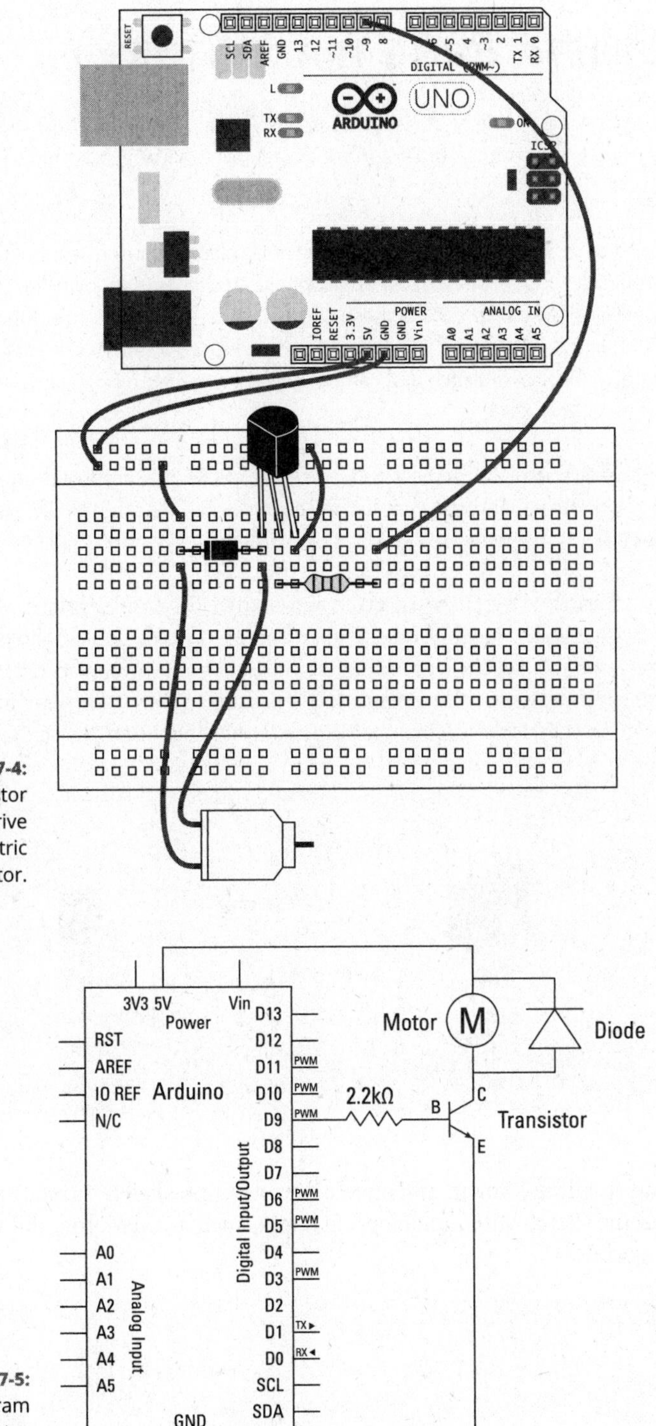

FIGURE 7-4: A transistor circuit to drive your electric motor.

FIGURE 7-5: A circuit diagram of a transistor circuit.

UNDERSTANDING TRANSISTORS

Sometime it's not possible or advisable to power an output directly from your Arduino pins. By using a transistor, you can control a bigger circuit from your modestly powerful Arduino.

Motors and other larger outputs (such as lots of LED lighting) often require more voltage and current than your Arduino pins can supply, so they need their own circuits to supply this power. To allow you to control these bigger circuits, you can use a component called a transistor. In the same way that a physical switch is used to turn a circuit on and off, a *transistor* is an electronic switch that can be turned on and off by using a very small voltage.

There are many transistors in the world, and each has its own product number that you can Google for details. The one I use in this section's example is a P2N2222A, which is an NPN-type transistor. (There are two kinds of transistor — NPN and PNP.)

A transistor has three legs: base, collector, and emitter. The *base* (or *gate*) is where the Arduino digital signal is sent; the *collector* (or *drain*) is the power source; and the *emitter* (or *source*) is the ground. The legs are numbered and, you hope, named in the datasheet to tell you which leg is which. In a circuit diagram, a transistor is drawn as in the following figure, with the collector at the top, the base to the left, and the emitter at the bottom.

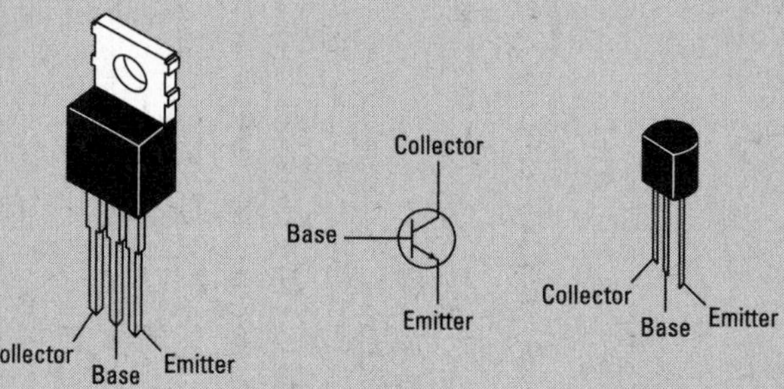

Build the circuit as shown, and open a new Arduino sketch. Press the Save button and save the sketch with a memorable name, such as myMotor, and then type the following code:

```
int motorPin = 9;

void setup() {

  pinMode(motorPin, OUTPUT);
```

```
}

void loop() {

  digitalWrite(motorPin, HIGH);
  delay(1000);

  digitalWrite(motorPin, LOW);
  delay(1000);

}
```

After you've typed the sketch, save it and press the Compile button to check your code. The Arduino IDE (introduced in Chapter 2) checks your code for any syntax errors (code grammar) and highlights them in the message area. The most common mistakes include typos, missing semicolons, and case sensitivity.

If the sketch compiles correctly, click Upload to upload the sketch to your board. You should see your motor spinning for one second and stopping for one second repeatedly.

If that's not what happens, you should double-check your wiring:

- Make sure that you're using pin number 9.
- Check that your diode is facing the correct way, with the band facing the 5v connection.
- Check the connections on the breadboard. If the jump wires or components are not connected using the correct rows in the breadboard, they will not work.

Understanding the Motor sketch

The Motor sketch is a basic sketch, and you may have noticed that it's a variation on the Blink sketch. This example changes the hardware but uses the same code to control an LED.

First, the pin is declared using digital pin 9:

```
int motorPin = 9;
```

In the `setup` function, pin 9 is defined as an output:

```
void setup() {

  pinMode(motorPin, OUTPUT);

}
```

The loop tells the output signal to go to `HIGH`, wait for 1000mS (1 second), go to `LOW`, wait for another 1000mS, and then repeat. This scenario gives you the most basic of motor control, telling the motor when to go on and off:

```
void loop() {

  digitalWrite(motorPin, HIGH);
  delay(1000);

  digitalWrite(motorPin, LOW);
  delay(1000);

}
```

Changing the Speed of Your Motor

On and off is all well and good, but sometimes you want greater control over the speed of your motor. The following sketch shows you how to control the speed of your motor with the same circuit.

The MotorSpeed sketch

Using the same circuit as in the preceding section, open a new Arduino sketch, save it with another memorable name, such as myMotorSpeed, and then type the following code:

```
int motorPin = 9;

void setup(){

  pinMode(motorPin, OUTPUT);

}
```

```
void loop() {

  for(int motorValue = 0; motorValue <= 255; motorValue +=5){
   analogWrite(motorPin, motorValue);
   delay(30);
  }

  for(int motorValue = 255; motorValue >= 0; motorValue -=5){
   analogWrite(motorPin, motorValue);
   delay(30);
  }

}
```

After you've typed the sketch, save it and press the Compile button to check your code. The Arduino IDE should highlight any grammatical errors in the message area if they are discovered.

If the sketch compiles correctly, click Upload to upload the sketch to your board. When uploading is finished, you should have a motor that spins very slowly to start, speeds up to its fastest spin, spins back down to a stop, and then repeats. It can be difficult to see this, so you should affix something more visible, such as a piece of tape or adhesive putty (such as Blu-Tack), to the motor spindle to see what's going on.

You may find that at its slowest point, the motor just hums. This hum is not a problem; it just means that the electromagnet doesn't have enough voltage to spin the motor; more voltage is required to generate the magnetism and gain momentum.

Understanding the MotorSpeed sketch

The MotorSpeed sketch is a slight variation of the Fade sketch described in Chapter 6, but works in the same way.

The pin you're using to control the motor circuit, digital pin 9, is declared:

```
int motorPin = 9;
```

Because it's an output, you define it in the `setup` function:

```
void setup() {
 pinMode(motorPin, OUTPUT);
}
```

In the main loop, you use analogWrite to send a PWM value to pin 9. This same principle is in the Fade sketch, which fades an LED. The first for loop sends a gradually increasing value to pin 9 until it reaches the maximum PWM value of 255. The second for loop gradually returns this value to 0; then the cycle repeats:

```
void loop() {

  for(int motorValue = 0; motorValue <= 255; motorValue +=5){
   analogWrite(motorPin, motorValue);
   delay(30);
  }

  for(int motorValue = 255; motorValue >= 0; motorValue -=5){
   analogWrite(motorPin, motorValue);
   delay(30);
  }

}
```

This process could be likened to revving a car engine. If you push down the pedal, you accelerate the engine to full speed. If you tap the gas pedal, the engine accelerates and then slows down. If you tap it at a constant rate before it slows, you maintain some of the momentum of the spinning motor and achieve an average (if somewhat jerky) speed. The transistor is doing this action, but very quickly. The intervals between on and off and the momentum of the motor allow you to achieve analog behavior from a digital signal.

Controlling the Speed of Your Motor

The sketch in the preceding section gave you control of the motor. In this section, you find out how to put some input into your circuit to give you full control of the motor on the fly.

The MotorControl sketch

To gain control of the speed of your motor whenever you need it, add a potentiometer to your circuit.

You need the following:

- An Arduino Uno
- A breadboard
- A transistor
- A DC motor
- A diode
- A 10k ohm variable resistor
- A 2.2k ohm resistor
- Jump wires

Follow the diagram in Figure 7-6 and the circuit diagram in Figure 7-7 to add a potentiometer alongside your motor control circuit.

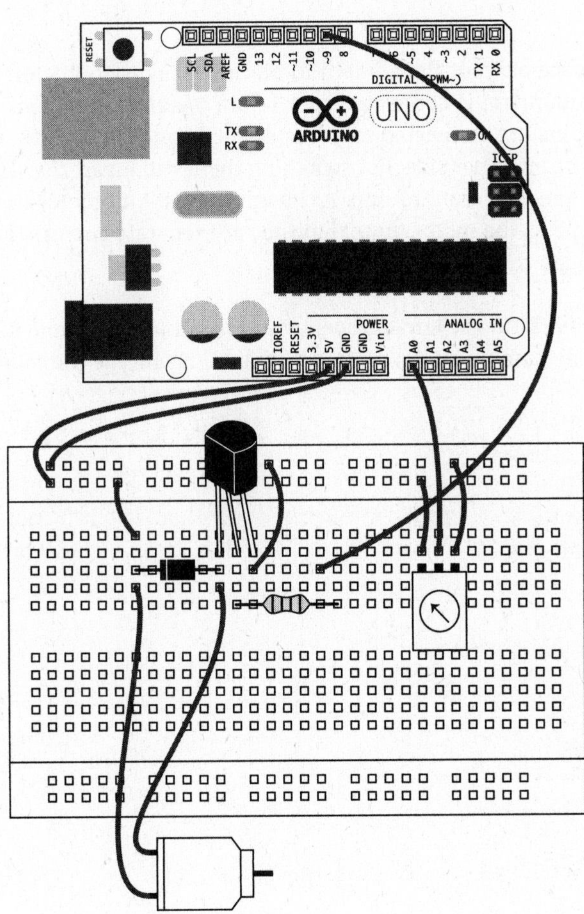

FIGURE 7-6: A transistor circuit to drive your electric motor.

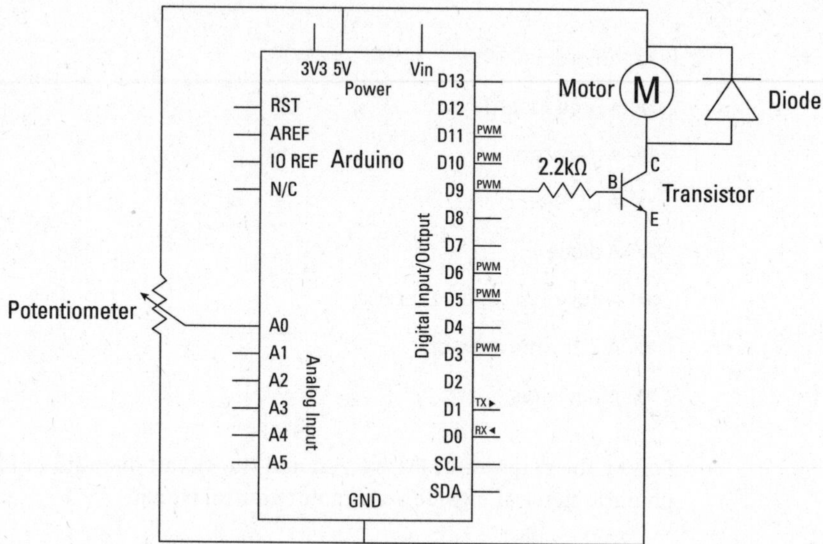

FIGURE 7-7: A circuit diagram of a transistor circuit.

Find a space on your breadboard to place your potentiometer. The central pin of the potentiometer is connected back to pin 9 using a jump wire, and the remaining two pins are connected to 5V on one side and GND on the other. The 5V and GND can be on either side, but switching them will invert the value that the potentiometer sends to the Arduino. Although the potentiometer uses the same power and ground as the motor, note that they are separate circuits that both communicate through the Arduino.

After you've built the circuit, open a new Arduino sketch and save it with another memorable name, such as myMotorControl. Then type the following code:

```
int potPin = A0;
int motorPin = 9;

int potValue = 0;
int motorValue = 0;

void setup() {
 Serial.begin(9600);
}

void loop() {
 potValue = analogRead(potPin);
 motorValue = map(potValue, 0, 1023, 0, 255);

 analogWrite(motorPin, motorValue);
```

```
  Serial.print("potentiometer = ");
  Serial.print(potValue);
  Serial.print("\t motor = ");
  Serial.println(motorValue);

  delay(2);
}
```

After you've typed the sketch, save it and click the Compile button to highlight any syntax errors.

If the sketch compiles correctly, click Upload to upload the sketch to your board. When it has finished uploading, you should be able to control your motor using the potentiometer. Turning the potentiometer in one direction speeds up the motor; turning it the other way slows down the motor. The next section explains how the code allows the potentiometer to change the speed.

Understanding the MotorControl Sketch

The MotorControl sketch is a variation on the AnalogInOutSerial sketch and works the same way, with a few name changes to better indicate what you are controlling and monitoring on the circuit.

As always, you declare the different variables used in the sketch. You use `potPin` to assign the potentiometer pin and `motorPin` to send a signal to the motor. The `potValue` variable is used to store the raw value of the potentiometer, and the `motorValue` variable stores the converted value that you want to output to the transistor to switch the motor:

```
int potPin = A0;
int motorPin = 9;

int potValue = 0;
int motorValue = 0;
```

For more details on the workings of this sketch, see the AnalogInOutSerial example in Chapter 6.

Tweaking the MotorControl sketch

You may find that the motor just hums below a minimum speed. It does so because it doesn't have enough power to spin. By monitoring the values sent to the motor using the MotorControl sketch, you can find the motor's minimum value to turn, and then optimize `motorValue` to turn the motor within its true range.

To find the range of `motorValue`, follow these steps:

1. **With the MotorControl sketch uploaded, click the serial monitor button at the top right of your Arduino window.**

   The serial monitor window displays the potentiometer value followed by the output value being sent to the motor, in this fashion:

   potentiometer = 1023   motor = 255

   These values are displayed in a long list and are updated as you turn the potentiometer. If you don't see the list scrolling down, make sure that the Autoscroll option is selected.

2. **Starting with your potentiometer reading a value of 0, turn your potentiometer very slowly until the humming stops and the motor starts spinning.**
3. **Make a note of the value displayed at this point.**
4. **Use an `if` statement to tell the motor to change speed only if the value is greater than the minimum speed needed to spin the motor, as follows:**

   a. *Find the part of your code that writes the `motorValue` to the motor:*

   ```
   analogWrite(motorPin, motorValue);
   ```

   b. *Replace it with the following piece of code:*

   ```
   if(motorValue > yourValue) {
     analogWrite(motorPin, motorValue);
   } else {
     digitalWrite(motorPin, LOW);
   }
   ```

5. **Replace `yourValue` with the number that you noted.**

   If the value is greater than `motorValue`, the motor speeds up. If it is lower, the pin is written `LOW` so that it is fully off. You could also type `analogWrite (motorPin, 0)` to accomplish the same thing. Tiny optimizations like this can help your project function smoothly, with no wasted movement or values.

Getting to Know Servo motors

A servo motor is made up of a motor and a device called an encoder that can track the rotation of the motor. *Servo motors* are used for precision movements, moving by a number of degrees to an exact location. Using your Arduino, you can tell the servo motor what degree you want it to move to, and it will go there from its current position. Most servo motors can move only 180 degrees, but you can use gearing to extend this range.

The servo in your kit will most likely be a hobby servo, similar to those shown in Figure 7-8. A hobby servo motor has plastic gears and can manage only relatively light loads. After you experiment with small servos, you have plenty of larger ones to choose from for heavy lifting. Servos are widely used in the robotics community for walkers that need precise movement in each of their feet.

FIGURE 7-8: A servo motor.

The examples in the following section walk you through the basic operations of sending signals to a servo and controlling one directly with a potentiometer.

Creating Sweeping Movements

This first servo motor example requires only a servo motor and will allow you to turn the motor through its full range of movement. The servo sweeps from 0° to 179° and then back again, in a similar way to the movement of an old rotary clock.

The Sweep sketch

You need the following for the Sweep sketch:

- An Arduino Uno
- A servo
- Jump wires

The wiring for a servo is simple because it comes with a neat three-pin socket. To connect it to your Arduino, simply use jump wires between the Arduino pins and the servo sockets directly or use a set of header pins to connect the socket to your breadboard.

As shown in Figures 7-9 and 7-10, the servo has a set of three sockets with wires connected to them, usually red, black, and white. All the calculations and readings to move the motor are done on the circuitry inside the servo itself, so all that is needed is power and a signal from the Arduino.

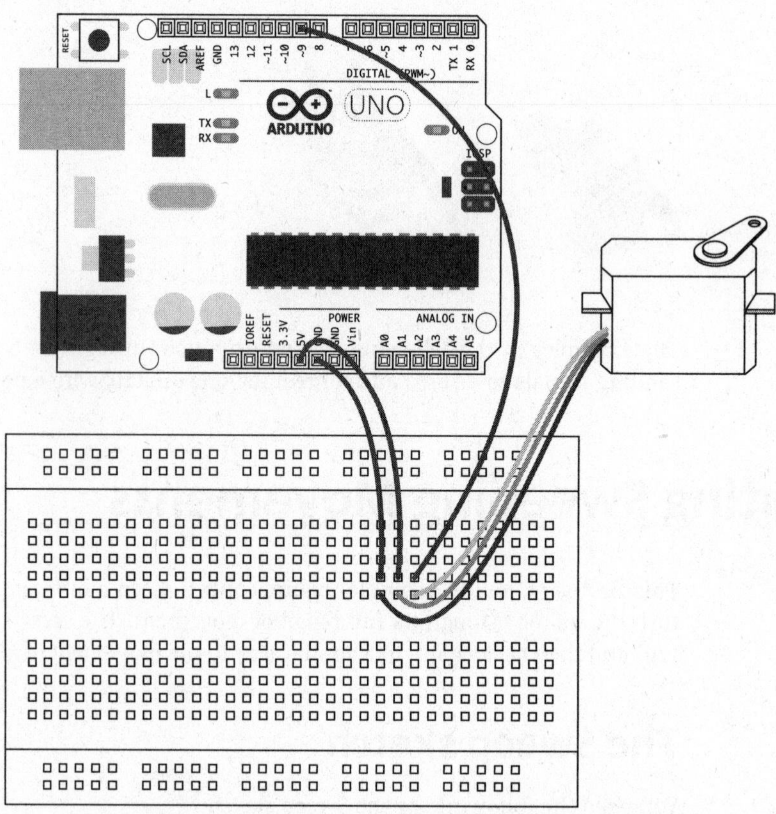

FIGURE 7-9: A servo motor wired to your Arduino.

Red is connected to 5V on the Arduino to power the motor and the circuitry inside it; black is connected to GND to ground the servo; and white is connected to pin 9 to control the servo's movement. The colors of these wires can vary, so always check the datasheet or any available documentation for your specific motor. Other common colors are red (5V), brown (GND), and yellow (signal).

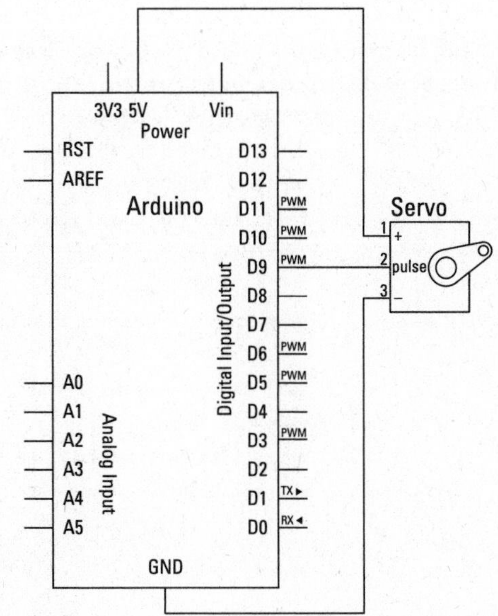

FIGURE 7-10: A circuit diagram of the servo circuit.

Complete the circuit as described and open the Sweep sketch by choosing File ➪ Examples ➪ Servo ➪ Sweep. The Sweep sketch is as follows:

```
/* Sweep
 by BARRAGAN <http://barraganstudio.com>
 This example code is in the public domain.

 modified 8 Nov 2013
 by Scott Fitzgerald
 http://www.arduino.cc/en/Tutorial/Sweep
*/


#include <Servo.h>

Servo myservo; // create servo object to control a servo
// twelve servo objects can be created on most boards

int pos = 0;   // variable to store the servo position

void setup() {
 myservo.attach(9); // attaches the servo on pin 9 to the servo object
}
```

```
void loop() {
 for(pos = 0; pos < 180; pos += 1) // goes from 0 degrees to 180 degrees
 {                                  // in steps of 1 degree
  myservo.write(pos);               // tell servo to go to position in
                                    // variable 'pos'
  delay(15);                        // waits 15ms for the servo to reach
                                    // the position
 }
 for(pos = 180; pos>=1; pos-=1)     // goes from 180 degrees to 0 degrees
 {
  myservo.write(pos);               // tell servo to go to position in
                                    // variable 'pos'
  delay(15);                        // waits 15ms for the servo to reach
                                    // the position
 }
}
```

After you find the sketch, press the Compile button to check the code. The compiler should, as always, highlight any grammatical errors in red in the message area.

If the sketch compiles correctly, click Upload to upload the sketch to your board. When the sketch has finished uploading, your motor should start turning backward and forward through 180 degrees, doing a dance on the table.

If nothing happens, you should double-check your wiring:

- Make sure that you're using pin 9 for the data (white/yellow) line.
- Check that you have the other servo wires connected to the correct pins.

Understanding the Sweep sketch

A servo library is included at the start of the Sweep sketch. This library will help you to get a lot out of your servo with little complex code:

```
#include <Servo.h>
```

The next line makes a servo object. The library knows how to use servos but needs you to give each one a name so that it can talk to each one. In this case, the new Servo object is called `myservo`. Using a name is similar to naming your variables; that is, they can be any name as long as they're consistent throughout your code and you don't use any Arduino reserved names, such as `int` or `delay`:

```
Servo myservo; // create servo object to control a servo
               // a maximum of eight servo objects can be created
```

The final line in the declarations is a variable to store the position of the servo:

```
int pos = 0;  // variable to store the servo position
```

In the `setup` function, the only item to set is the pin number of the Arduino pin that is communicating with the servo. In this case, you are using pin 9, but it could be any PWM pin.

```
void setup()
{
  myservo.attach(9); // attaches the servo on pin 9 to the servo object
}
```

The `loop` function performs two simple actions, and both are `for` loops. The first `for` loop gradually increases the `pos` variable from 0 to 180. Because of the library, you can write values in degrees rather than the normal 0 to 255 used for PWM control. With every loop, the value is increased by 1 and sent to the servo using a function specific to the servo library, `<servoName>.write(<value>)`. After the loop updates the value, a short delay of 15 milliseconds occurs while the servo reaches its new location.

REMEMBER

In contrast to other outputs, after a servo is updated, it starts moving to its new position instead of needing to be told to do so.

```
void loop()
{
  for(pos = 0; pos < 180; pos += 1) // goes from 0 degrees to 180 degrees
  {                                 // in steps of 1 degree
    myservo.write(pos);             // tell servo to go to position in
                                    // variable 'pos'
    delay(15);                      // waits 15ms for the servo to reach
                                    // the position
  }
```

The second `for` loop does the same in the opposite direction, returning the servo to its start position:

```
  for(pos = 180; pos>=1; pos-=1)   // goes from 180 degrees to 0 degrees
  {
    myservo.write(pos);            // tell servo to go to position in
                                   // variable 'pos'
```

```
    delay(15);                       // waits 15ms for the servo to reach
                                     // the position
  }
}
```

It's a good idea to use this simple servo example to test whether your servo is working correctly, before coding more complex examples.

Controlling Your Servo

Now that you have mastered control of the servo, you can try something with a bit more interaction. By using a potentiometer (or any analog sensor), you can directly control your servo in the same way that you'd control a mechanical claw at the arcades.

The Knob sketch

The Knob example shows you how you can easily use a potentiometer to move your servo to a specific degree.

You need the following:

- An Arduino Uno
- A breadboard
- A servo
- A 10k ohm variable resistor
- Jump wires

The servo is wired exactly as in the Sweep example, but this time you need extra connections to 5V and GND for the potentiometer, so you must use a breadboard to provide the extra pins. Connect the 5V and GND pins on the Arduino to the positive (+) and negative (–) rows on the breadboard. Connect the servo to the breadboard by using either a row of three header pins or three jump wires.

Connect the red socket to the 5V row, the black/brown socket to the GND row, and the white/yellow socket to pin 9 on the Arduino. Find a space on the breadboard for the potentiometer. Connect the center pin to pin A0 on the Arduino and the remaining pins to 5V on one side and GND on the other. Refer to the circuit diagram in Figure 7-11 and the schematic in Figure 7-12.

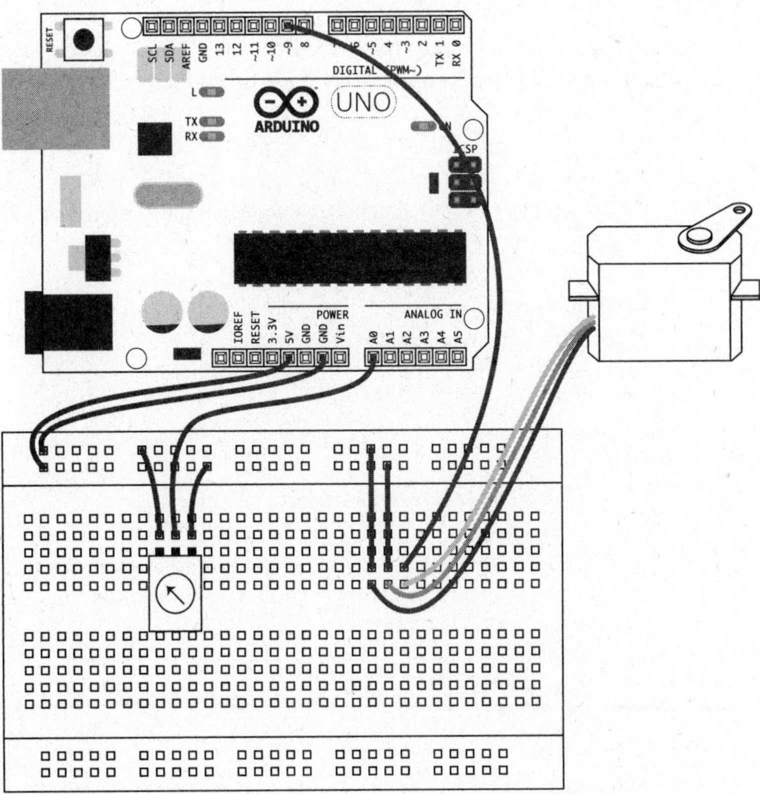

FIGURE 7-11: A servo motor with a control knob.

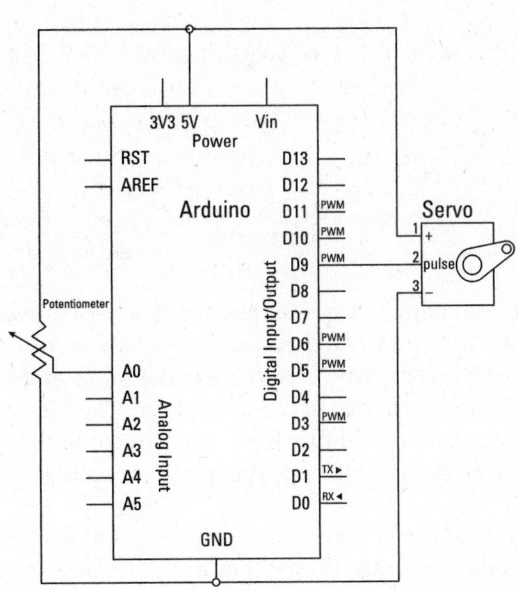

FIGURE 7-12: A circuit diagram for a servo and a potentiometer.

After you have built the circuit, open the sketch by choosing File ➪ Examples ➪ Servo ➪ Knob. The code for the sketch is as follows:

```
/*
 Controlling a servo position using a potentiometer (variable resistor)
 by Michal Rinott <http://people.interaction-ivrea.it/m.rinott>

 modified on 8 Nov 2013
 by Scott Fitzgerald
 http://www.arduino.cc/en/Tutorial/Knob
*/

#include <Servo.h>

Servo myservo; // create servo object to control a servo

int potpin = 0; // analog pin used to connect the potentiometer
int val;  // variable to read the value from the analog pin

void setup() {
 myservo.attach(9); // attaches the servo on pin 9 to the servo object
}

void loop() {
 val = analogRead(potpin);            // reads the value of the
                                      // potentiometer (value between
                                      // 0 and 1023)
 val = map(val, 0, 1023, 0, 179);     // scale it to use it with the
                                      // servo (value between 0 and 180)
 myservo.write(val);                  // sets the servo position according
                                      // to the scaled value
 delay(15);                           // waits for the servo to get there
}
```

TIP

With all Arduino examples, it's best to assume that they are works-in-progress and may not always be accurate. You may have noticed a few discrepancies between the comments and the code. When referring to the range of degrees to move the servo, the sketch mentions both 0 to 179 and 0 to 180. The correct range is 0 to 179, which gives you 180 values. Counting from zero is referred to as *zero indexing* and is a common occurrence in Arduino, as you may have noticed by this point.

After you've found the sketch, press the Compile button to check the code. If the compiler encounters any syntax errors, they are highlighted in the message area, which lights up red.

If the sketch compiles correctly, click Upload to upload the sketch to your board. When it is has finished uploading, your servo should turn as you turn your potentiometer.

If that isn't what happens, you should double-check your wiring:

- Make sure that you're using pin 9 to connect the data (white/yellow) line to the servo.
- Check your connections to the potentiometer and make sure that the center pin is connected to analog pin 0.
- Check the connections on the breadboard. If the jump wires or components are not connected using the correct rows in the breadboard, they will not work.

Understanding the Knob sketch

In the declarations, the servo library, Servo.h, and a new servo object are named. The analog input pin is declared with a value of 0, showing that you are using analog 0.

TIP

You may have noticed that the pin is numbered 0, not A0 as in other examples. Either is fine, because A0 is just an alias of 0, as A1 is of 1, and so on. Using A0 is good for clarity but optional.

One last variable stores the value of the reading, which will become the output:

```
#include <Servo.h>

Servo myservo; // create servo object to control a servo

int potpin = 0; // analog pin used to connect the potentiometer
int val;   // variable to read the value from the analog pin
```

In the `setup` function, the only item to define is `myservo`, which is using pin 9.

```
void setup()
{
  myservo.attach(9); // attaches the servo on pin 9 to the servo object
}
```

Rather than use two separate variables for input and output, the Knob sketch simply uses one. First, `val` is used to store the raw sensor data, a value from 0 to 1023. The `map` function processes this value by scaling its range to that of the servo: 0 to 179. `myservo.write` then writes this value to the servo. There is a 15-millisecond delay to allow the servo motor to reach its destination. Then the loop repeats and updates the position of the servo as necessary:

```
void loop()
{
  val = analogRead(potpin);            // reads the value of the potentiometer
                                       // (value between 0 and 1023)
  val = map(val, 0, 1023, 0, 179);     // scale it to use it with the servo
                                       // (value between 0 and 180)
  myservo.write(val);                  // sets the servo position according
                                       // to the scaled value
  delay(15);                           // waits for the servo to get there
}
```

With this simple addition to the circuit, it's possible to control a servo with any sort of input. In this example, the code uses an analog input, but with a few changes it could just as easily use a digital input.

Making Noises

If you've just finished the motor sketches, you have mastered movement and must be ready for a new challenge. In this section, you look at a project that's a bit more tuneful than the previous ones: making music (or noise at least) with your Arduino. Yes, you can make electronic music — albeit simple — using a piezo buzzer.

Piezo buzzer

A piezo or piezoelectric buzzer is found in hundreds of thousands of devices. If you hear a tick, buzz, or beep, it's likely caused by a piezo. The *piezo* is composed of two layers, a ceramic and a metal plate joined together. When electricity passes from one layer to the other, the piezo bends on a microscopic level and makes a sound, as shown in Figure 7-13.

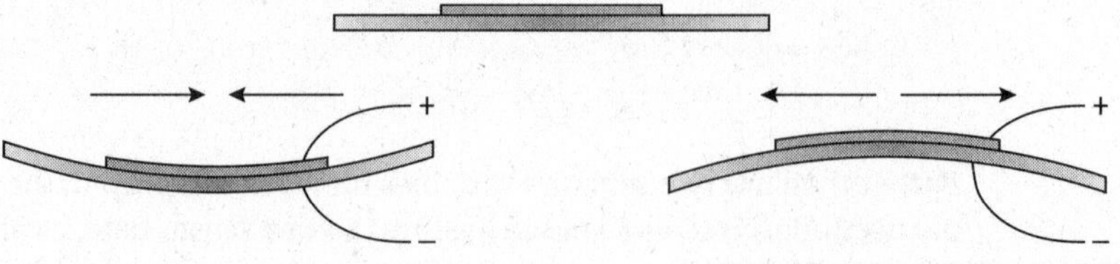

FIGURE 7-13: An exaggeration of the miniature movements of a piezo.

If you switch between a voltage and ground, the piezo bends and generates a tick sound; if this happens fast enough, these ticks turn into a tone. This tone can be quite harsh, similar to the old mobile phone ringtone or computer game sounds from the 1980s, and is known as a square wave. Every time the piezo changes polarity fully, it produces a square wave with abrupt, hard edges, like a square. Other types of waves include triangle waves and sine waves, which are progressively less harsh. Figure 7-14 is an illustration of these waves so you can see the differences between them.

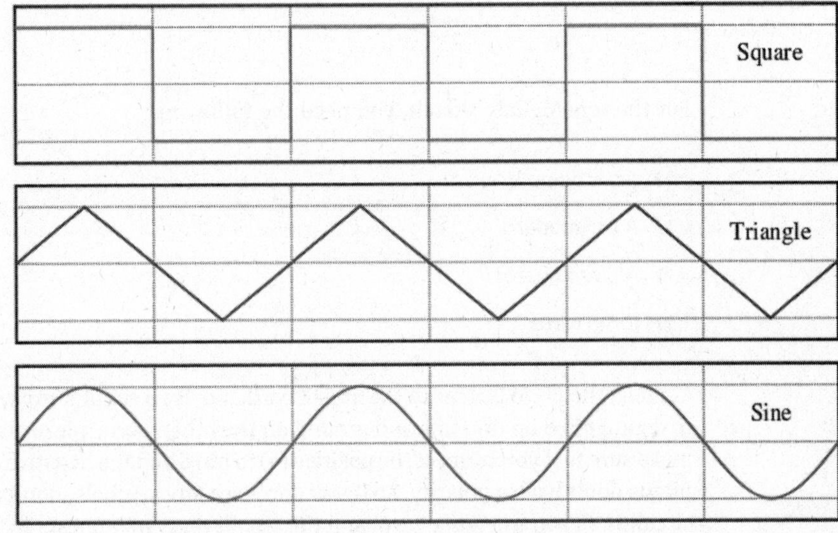

FIGURE 7-14: Square, triangle, and sine waves have different shapes, which produce different sounds.

Original illustration by Omegatron

Piezos generate square waves, resulting in a buzzing sound. The buzzer isn't restricted to just one pitch. By changing the frequency that the buzzer is switched at (the width between the square waves), you can generate different frequencies and therefore different notes.

The toneMelody sketch

With the toneMelody sketch, you see how to change the frequency of your piezo and play a predefined melody. This circuit allows you to program your own sounds. With a bit of time and consideration you can turn these sounds into melodies.

Piezo buzzers are supplied in most Arduino kits but can take many different forms. They can be supplied without an enclosure, as shown in Figure 7-15, or can be enclosed in plastic housing ranging from small cylinders to flat coin-like shapes.

They may also have different connections, either a set of two pins protruding from the underside of the piezo or two wires protruding from its side.

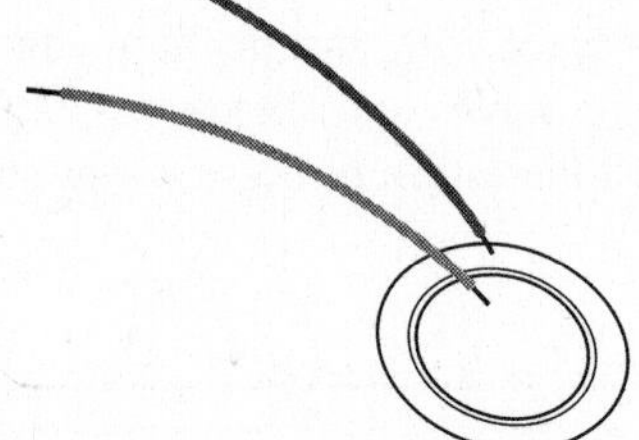

FIGURE 7-15: A piezo buzzer out of its enclosure.

For the toneMelody sketch, you need the following:

- An Arduino Uno
- A breadboard
- A piezo buzzer
- Jump wires

Connect the piezo buzzer to the breadboard and use a set of jump wires to connect it to digital pin 8 on one side and ground on the other. Some piezos have a polarity, so make sure that you connect the positive (+) to pin 8 and the negative (–) to GND. Other piezos don't have a polarity, so if you don't see any symbols, don't worry. The piezo circuit is shown in Figure 7-16, and the circuit diagram appears in 7-17.

Complete the circuit and open the sketch by choosing File ⇨ Examples ⇨ 02. digital ⇨ toneMelody. You see the following code:

```
/*
  Melody

 Plays a melody

 circuit:
 * 8-ohm speaker on digital pin 8

 created 21 Jan 2010
 modified 30 Aug 2011
 by Tom Igoe
```

```
 This example code is in the public domain.

 http://arduino.cc/en/Tutorial/Tone

 */
 #include "pitches.h"

// notes in the melody:
int melody[] = {
  NOTE_C4, NOTE_G3,NOTE_G3, NOTE_A3, NOTE_G3,0, NOTE_B3, NOTE_C4};

// note durations: 4 = quarter note, 8 = eighth note, etc.:
int noteDurations[] = {
  4, 8, 8, 4,4,4,4,4 };

void setup() {
  // iterate over the notes of the melody:
  for (int thisNote = 0; thisNote < 8; thisNote++) {

    // to calculate the note duration, take one second
    // divided by the note type.
    //e.g. quarter note = 1000 / 4, eighth note = 1000/8, etc.
    int noteDuration = 1000/noteDurations[thisNote];
    tone(8, melody[thisNote],noteDuration);

    // to distinguish the notes, set a minimum time between them.
    // the note's duration + 30% seems to work well:
    int pauseBetweenNotes = noteDuration * 1.30;
    delay(pauseBetweenNotes);
    // stop the tone playing:
    noTone(8);
  }
}

void loop() {
  // no need to repeat the melody.
}
```

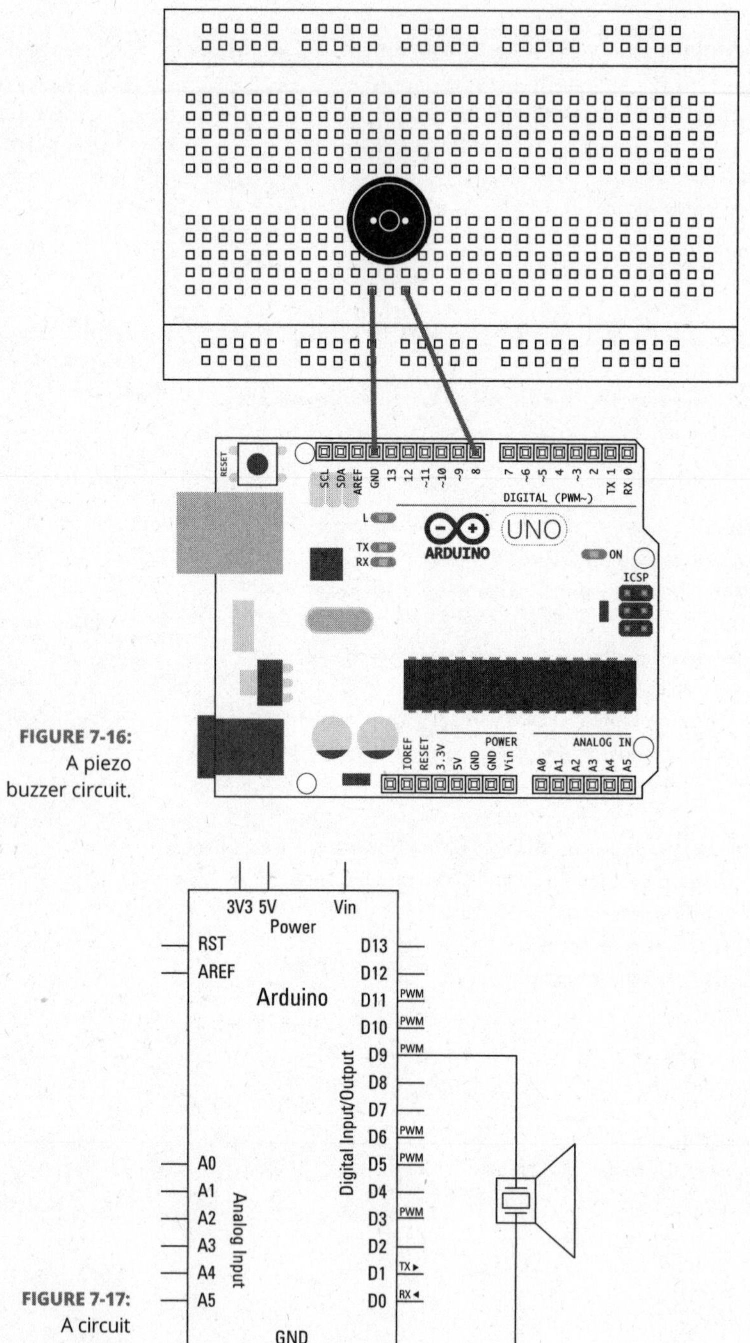

FIGURE 7-16: A piezo buzzer circuit.

FIGURE 7-17: A circuit diagram of a piezo buzzer circuit.

In the toneMelody sketch, you have another tab called pitches.h, which contains all the data needed to make the correct tones with your buzzer. In your Arduino sketch folder, this tab (and other, additional tabs) appears as its own individual file and must be included in the main sketch using the `#include` function followed by the name of the file to be included. In this case, the code reads `#include "pitches.h"`. The pitches.h file follows for your reference.

pitches.h

```
/*************************************************
 * Public Constants
 *************************************************/
#define NOTE_B0  31
#define NOTE_C1  33
#define NOTE_CS1 35
#define NOTE_D1  37
#define NOTE_DS1 39
#define NOTE_E1  41
#define NOTE_F1  44
#define NOTE_FS1 46
#define NOTE_G1  49
#define NOTE_GS1 52
#define NOTE_A1  55
#define NOTE_AS1 58
#define NOTE_B1  62
#define NOTE_C2  65
#define NOTE_CS2 69
#define NOTE_D2  73
#define NOTE_DS2 78
#define NOTE_E2  82
#define NOTE_F2  87
#define NOTE_FS2 93
#define NOTE_G2  98
#define NOTE_GS2 104
#define NOTE_A2  110
#define NOTE_AS2 117
#define NOTE_B2  123
#define NOTE_C3  131
#define NOTE_CS3 139
#define NOTE_D3  147
#define NOTE_DS3 156
#define NOTE_E3  165
#define NOTE_F3  175
#define NOTE_FS3 185
#define NOTE_G3  196
#define NOTE_GS3 208
```

(continued)

```
#define NOTE_A3  220
#define NOTE_AS3 233
#define NOTE_B3  247
#define NOTE_C4  262
#define NOTE_CS4 277
#define NOTE_D4  294
#define NOTE_DS4 311
#define NOTE_E4  330
#define NOTE_F4  349
#define NOTE_FS4 370
#define NOTE_G4  392
#define NOTE_GS4 415
#define NOTE_A4  440
#define NOTE_AS4 466
#define NOTE_B4  494
#define NOTE_C5  523
#define NOTE_CS5 554
#define NOTE_D5  587
#define NOTE_DS5 622
#define NOTE_E5  659
#define NOTE_F5  698
#define NOTE_FS5 740
#define NOTE_G5  784
#define NOTE_GS5 831
#define NOTE_A5  880
#define NOTE_AS5 932
#define NOTE_B5  988
#define NOTE_C6  1047
#define NOTE_CS6 1109
#define NOTE_D6  1175
#define NOTE_DS6 1245
#define NOTE_E6  1319
#define NOTE_F6  1397
#define NOTE_FS6 1480
#define NOTE_G6  1568
#define NOTE_GS6 1661
#define NOTE_A6  1760
#define NOTE_AS6 1865
#define NOTE_B6  1976
#define NOTE_C7  2093
#define NOTE_CS7 2217
#define NOTE_D7  2349
#define NOTE_DS7 2489
#define NOTE_E7  2637
#define NOTE_F7  2794
#define NOTE_FS7 2960
#define NOTE_G7  3136
```

```
#define NOTE_GS7 3322
#define NOTE_A7  3520
#define NOTE_AS7 3729
#define NOTE_B7  3951
#define NOTE_C8  4186
#define NOTE_CS8 4435
#define NOTE_D8  4699
#define NOTE_DS8 4978
```

After you've found the sketch, press the Compile button to check the code. The message area highlights grammatical errors in red if any are discovered.

If the sketch compiles correctly, press Upload to upload the sketch to your board. When it has finished uploading, you should hear a buzzer that sings a tune to you and then stops. To hear the tune again, press the reset button on your Arduino.

If you don't hear a buzzer, you should double-check your wiring:

- » Make sure you're using pin 8 as your output.
- » Check that your piezo is correctly positioned. Symbols may be hidden on the underside if they are not visible on the top. If you don't see any markings, try the piezo in the other orientation.
- » Check the connections on the breadboard. If the jump wires or components are not connected using the correct rows in the breadboard, they will not work.

Understanding the sketch

The toneMelody sketch is the first one in this book that uses multiple tabs. You sometimes use *tabs* as a convenient way of separating sketches. In this case, the `pitches.h` tab is a reference or lookup table for all the possible notes in the piezo's range. Because this code won't change, it doesn't need to be in the main body of code.

At the top of the toneMelody sketch is a note to include `pitches.h`, which is treated in the same way as a library. It is an external file that can be brought into sketches if needed. In this case, we need it to determine which frequencies are used to create the notes:

```
#include "pitches.h"
```

INTRODUCING ARRAYS

In its simplest form, an *array* is a list of data. Think of it as being like a shopping list, as shown in the following table. Each row has a number, referred to as the *index*, and the data contained in that part of the list. This kind of array is a one-dimensional array, containing only one item of data for each, in this case the name of a piece of fruit.

In computing, it's common to use *zero indexing*, which means starting your list from zero, as shown in the following table.

Index	Value
0	apples
1	bananas
2	oranges

How is an array relevant to Arduino? Arrays can store integers, floats, characters, or any other type of data, but I use integers here to keep things simple. Here is an array of six integer values:

```
int simpleArray[] =
       {1, 255, -51, 0, 102, 27};
```

First, `int` defines the type of data being stored as integers (whole numbers). The data type could also be `float` for floating-point numbers or `char` for characters. The name of the array is `simpleArray` but can be any relevant name that best describes your array. The square brackets ([ ]) store the length of the array (the number of values that can be stored in the array); in this case, the space is blank, which means that this array has no fixed length. The numbers inside the curly braces { } are values defined in the array. These are optional, so if they are not defined, the array will be left empty.

There are other correct ways to declare arrays, including the following:

```
int simpleArray[10];
float simpleArray[5] = {2.7, 42.1, -9.1, 300.6};
char simpleArray[14] = "hello, world!";
```

Note that the character array entry has a number one greater than the number of characters. Remember this requirement if you're getting errors.

Now that your array is defined, you need to know how to use it. To use values in an array, you refer to them by their index. If you wanted to send a value to the serial monitor, you would write the following:

```
Serial.println(simpleArray[2]);
```

This line would display a value of –51 because that is the value stored in index 2 of the array.

You can also update values in the array. An effective way to update is with a `for` loop, to count through each index in the array (see Chapter 10 for more details) as in the following example:

```
for (int i = 0; i < 6; i++) {
    simpleArray[i] = analogRead(sensorPin);
}
```

The `for` loop in this case will loop six times, increasing the `i` variable by 1 each loop. The `i` variable is used also to represent the index of the array, so with each loop, a new analog reading from `sensorPin` is stored in the current index and the index of the array is incremented for the next loop.

This loop is a clever and efficient way to work through arrays, either using or updating the data stored in them. Arrays can get even more complicated, storing many strings of text, or they can be multidimensional, like a spreadsheet, with many values associated with each index. For more information, head over to the official Arduino reference page on arrays at `http://arduino.cc/en/Reference/Array`.

Now that the sketch knows the different notes, the melody is defined in an array so that the notes can be stepped through in order. To find out more about arrays, see the "Introducing Arrays" sidebar. The names, such as `NOTE_C4`, refer to the names of notes in the pitches.h tab. If you look at pitches.h, you will see that it uses a C function called `define` for each of these note references and follows them with a number, such as `#define NOTE_C4 262`. So whenever `NOTE_C4` is mentioned, it is really just a variable name for the value 262:

```
// notes in the melody:
int melody[] = {
  NOTE_C4, NOTE_G3,NOTE_G3, NOTE_A3, NOTE_G3,0, NOTE_B3, NOTE_C4};
```

Without the beat, your melody wouldn't sound right, so another array stores the duration for each note:

```
// note durations: 4 = quarter note, 8 = eighth note, etc.:
int noteDurations[] = {
  4, 8, 8, 4,4,4,4,4 };
```

In setup, a for loop is used to cycle through each of the eight notes, from 0 to 7. The thisNote value is used as an index to point to the correct items in each array:

```
void setup() {
  // iterate over the notes of the melody:
  for (int thisNote = 0; thisNote < 8; thisNote++) {
```

The duration is calculated by dividing 1,000 (or 1 second) by the required duration, 4 for a quarter note or crotchet, 8 for and eighth note or quaver, and so on. This value is then written to the function tone, which sends the current note to pin 8 for the assigned duration:

```
    // to calculate the note duration, take one second
    // divided by the note type.
    //e.g. quarter note = 1000 / 4, eighth note = 1000/8, etc.
    int noteDuration = 1000/noteDurations[thisNote];
    tone(8, melody[thisNote],noteDuration);
```

A small pause between notes is used so that they are better defined. In this case, the pause is relative to the length of the note and is set to 30 percent of the current note duration:

```
// to distinguish the notes, set a minimum time between them.
    // the note's duration + 30% seems to work well:
    int pauseBetweenNotes = noteDuration * 1.30;
    delay(pauseBetweenNotes);
```

Next, the noTone function is used to turn off pin 8, stopping the note after it has played for its duration:

```
    // stop the tone playing:
   noTone(8);
   }
 }
```

In the loop, nothing happens. As it stands, the melody plays once at the start and then ends. The melody could be moved to the loop to play forever, but this decision may cause mild headaches:

```
void loop() {
  // no need to repeat the melody.
}
```

TIP

The toneMelody sketch is a great example of using a melody as an audio signal at the start of a sketch. Audio feedback can be a great alternative or addition to visual feedback in your project.

Making an Instrument

In the preceding section, you find out how to make your project play a sound rather than blink a light, as in previous sketches. In the example in this section, you see how to go beyond playing a sound — you create your own instrument, similar to the Theremin. The *Theremin*, named after its inventor Léon Theremin, was one of the first electronic instruments, developed in the 1920s. It worked by detecting the electromagnetic field of the player's hands to change signals: one hand for volume and the other for pitch.

The PitchFollower sketch

In the PitchFollower sketch, you find out how to make a budget Theremin by using a piezo as a light sensor to control the pitch.

You need the following:

- An Arduino Uno
- A breadboard
- A piezo
- A light sensor
- A 4.7k ohm resistor
- Jump wires

This circuit has two separate halves: the piezo and the light sensor circuit. The piezo is wired as in the toneMelody sketch, with one wire to digital pin 9 and the other to GND. The light sensor is connected to analog 0 on one side and 5V on the other; the 4.7K resistor is connected between analog 0 and ground (as shown in Figures 7-18 and 7-19). If you do not have a 4.7K resistor, use the nearest you have to that value.

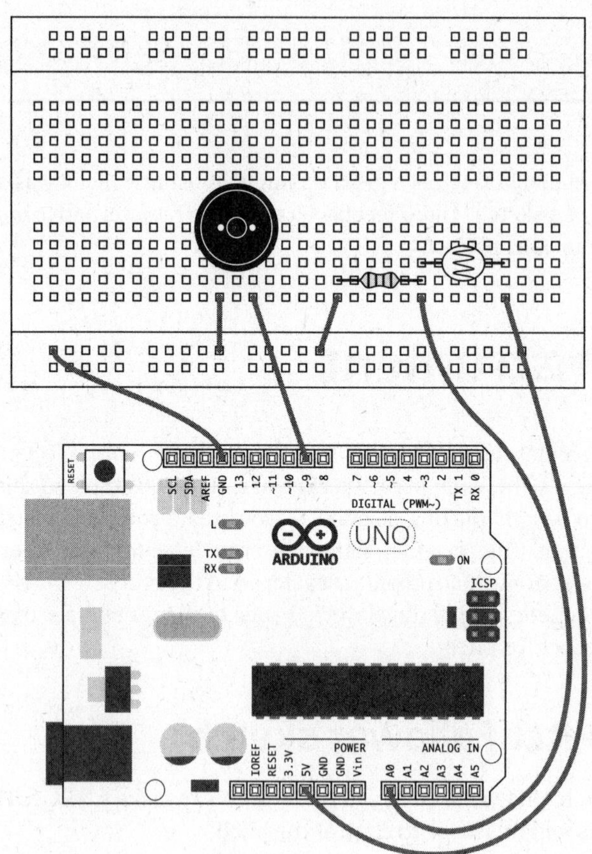

FIGURE 7-18: A light-sensor-controlled Theremin circuit.

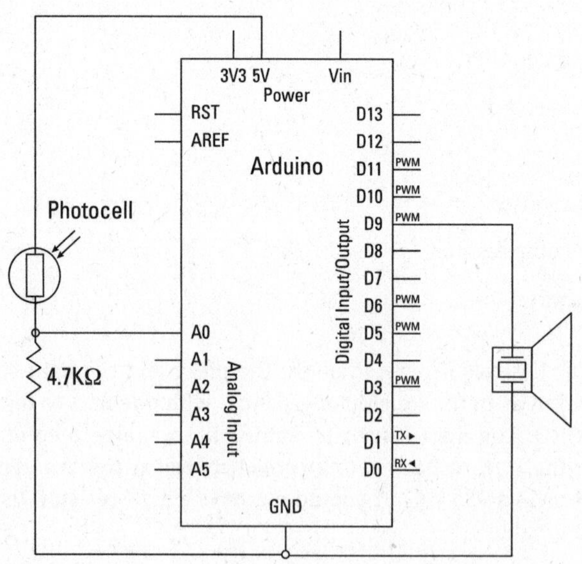

FIGURE 7-19: A circuit diagram of the light-sensor-controlled Theremin.

Complete the circuit, and open the sketch by choosing File ➪ Examples ➪ 02. Digital ➪ tonePitchFollower:

```
/*
  Pitch follower

 Plays a pitch that changes based on a changing analog input

 circuit:
 * 8-ohm speaker on digital pin 9
 * photoresistor on analog 0 to 5V
 * 4.7K resistor on analog 0 to ground

 created 21 Jan 2010
 modified 9 Apr 2012
 by Tom Igoe

This example code is in the public domain.

 http://arduino.cc/en/Tutorial/Tone2

 */

void setup() {
  // initialize serial communications (for debugging only):
  Serial.begin(9600);
}

void loop() {
  // read the sensor:
  int sensorReading = analogRead(A0);
  // print the sensor reading so you know its range
  Serial.println(sensorReading);
  // map the analog input range (in this case, 400 - 1000 from the
  // photoresistor) to the output pitch range (120 - 1500Hz)
  // change the minimum and maximum input numbers below depending on the
  // range your sensor's giving:
int thisPitch = map(sensorReading, 400, 1000, 120, 1500);

  // play the pitch:
  tone(9, thisPitch, 10);
  delay(1);        // delay in between reads for stability
}
```

After you've found the sketch, press the Compile button to check the code. Any syntax errors turn the message area red when they are discovered, and you see an error message stating what is wrong.

If the sketch compiles correctly, press Upload to upload the sketch to your board. When it has finished uploading, you should have a light sensor that will change the pitch of your buzzer. If you don't hear a change, make sure that you are in a well-lit area or turn a desk lamp on over your breadboard to increase the difference when you cover the light sensor with your hand.

If nothing happens, you should double-check your wiring:

- Make sure that you're using the correct pin number for the inputs and outputs.
- Check that your piezo is turned the correct way. Symbols may be hidden on the underside if they are not visible.
- Check the connections on the breadboard. If the jump wires or components are not connected using the correct rows in the breadboard, they will not work.

Understanding the sketch

This sketch is a lot shorter than the toneMelody sketch, presented earlier in the chapter, because it converts the readings from the light sensor to a frequency directly rather than requires a lookup table. Because you are converting readings, you can slide between notes as well as choose them individually.

In the `setup` function, the serial port is opened to allow you to monitor the sensor readings as they come in:

```
void setup() {
  // initialize serial communications (for debugging only):
  Serial.begin(9600);
}
```

In the main loop, the light sensor is read from analog pin 0. This reading is also forwarded to the serial monitor:

```
void loop() {
  // read the sensor:
  int sensorReading = analogRead(A0);
  // print the sensor reading so you know its range
  Serial.println(sensorReading);
```

To convert the sensor's range to the range of frequencies that the buzzer can cover, you use the map function:

```
// map the analog input range (in this case, 400 - 1000 from the
// photoresistor) to the output pitch range (120 - 1500Hz)
// change the minimum and maximum input numbers below depending on the
// range your sensor's giving:
int thisPitch = ma(sensorReading, 400, 1000, 100, 1000);
```

The tone function then outputs the note with the mapped sensor value and a very short duration of 10 milliseconds. This duration serves to make the sound audible, but the real duration is determined by how long you hold your hand over the sensor, as described previously:

```
// play the pitch:
tone(9, thisPitch, 10);
```

Finally, a tiny delay occurs at the end of the loop to improve the stability for the readings:

```
  delay(1);        // delay in between reads for stability
}
```

With this setup, you can quickly make an easy controller and maybe even form a traveling Theremin band with your friends.

3 Building on the Basics

IN THIS PART . . .

See some of the varied uses of Arduino though a few projects already out in the world.

Make your basic prototypes into something more solid by learning about soldering.

Use code to improve the reliability of your project.

Learn how to choose the right sensors for your projects.

Discover new possibilities by using shields and libraries.

» Understanding how these projects work and why they're so great

» Getting ideas for your own amazing Arduino projects

Chapter 8
Learning by Example

In previous chapters, I show you the beginnings of Arduino projects, but it can be difficult to know what to do with that knowledge. Read this chapter for a taste of the vast variety of uses and huge potential of Arduino. You discover projects that have used Arduino to create breathtaking art installations, long-lasting interactive exhibitions, and prototypes for products that have made it out into the world.

Skube

The Skube project was developed by Andrew Nip, Ruben van der Vleuten, Malthe Borch, and Andrew Spitz as part of the Tangible User Interface module at the Copenhagen Institute of Interaction Design (CIID). It's an excellent example of how to use Arduino for product prototyping and development.

Skube (shown in Figure 8-1) is a product that allows you to interact with digital music services that you would usually access on your computer. To make better use of digital music services, the project aims to rethink the way audio devices work and how they are used. Each Skube has two modes, Playlist and Discovery, that you select by tapping the top of the device. Playlist plays through a predefined playlist of music, and Discovery searches for similar artists or tracks. Skubes can also be combined to shuffle between each of the predefined playlists. They will physically snap together, giving the user a tangible way of mixing different playlists and discovering new music.

FIGURE 8-1: A single Skube.

Courtesy of Andrew Spitz

How it works

Thankfully, the Skube team offers a great amount of documentation and an excellent video of the finished prototypes as well as a good look under the hood. Inside each Skube is an Arduino and an XBee wireless module (unfortunately, this book doesn't have enough room to cover this amazing module, but you should find ample resources online with a quick Google search).

The Arduino's main function is to act as the middleman, hosting a number of different sensors and communication devices and relaying the correct data to the correct places. The tap sensor is the same as that described in Chapter 11 and uses a simple piezo element to monitor vibrations. When Skubes snap together, the magnet is acting as a switch as well, activating a reed switch. The *reed switch* closes its metal contacts when a magnet comes near, giving the device a clear indication of when the magnet is present.

The project also has an FM radio module, which is used to output the music for each Skube. With the XBee wireless modules, you can communicate with and coordinate each Skube by using custom software on a computer, which is written in Max/MSP language, a visual programming language found in many audio and music projects.

The Skube team gathered data from playlists on Last.fm and Spotify to find similar artist features. These companies provide various features to their customers (such as assembling playlists based on your favorite tracks and providing a database of albums and artists), as well as access to these features for developers who have interesting ideas for projects, smartphone applications, or products. This resource is known as an application programming interface (API). The API for Last.fm (`www.last.fm/api`) and Spotify (`developer.spotify.com/technologies/web-api/`) are just two examples; many more are available for other specific web services as well, so always give it a Google!

You can see that this project has many elements, involving not just communicating wirelessly using Arduino (which can be a task in itself) but also communicating with other software, and from that software, communicating with other services on the Internet. For the more code-savvy among you, this application of Arduino may allow you to build from your existing knowledge to integrate hardware with other software. In Part 5, I introduce other available software and look in more depth at communicating with the Processing software environment.

Further reading

You can find out more about this project by checking out the project pages on the CIID website at `ciid.dk/education/portfolio/idp12/courses/tangible-user-interface/projects/skube/` and Andrew Spitz's website at `www.soundplusdesign.com/?p=5516`.

Chorus

Chorus is a kinetic installation by United Visual Artists (UVA), a London-based art and design practice. UVA's work crosses many disciplines, including sculpture, architecture, live performance, moving image, and installation. The group has a reputation for creating visually stunning projects that push the boundaries of those disciplines.

Chorus (shown in Figure 8-2) uses sound, light, and motion with dramatic effect and is an excellent example of how Arduino can play a role in huge installations as well as in tiny prototypes. The installation is made up of eight tall black pendulums swinging back and forth, simultaneously emitting light and sound. Spectators can walk underneath the pendulums, immersing themselves in the performance, as "the rhythm of the work moves between chaos and unison." Each pendulum has its own score, composed by Mira Calix, which can be heard when the audience gets near enough.

FIGURE 8-2: Chorus in full swing.

Courtesy of United Visual Artists

How it works

In this project, an Arduino is used for not only light and sound but also the motion of the pendulums. The swing of each pendulum is controlled by an electric motor mounted on a reduction gearbox. This motor can be controlled with a relay circuit, allowing the Arduino to affect this huge mechanical object. Each pendulum has two custom circuit boards, each with 50 LEDs to provide the light for the piece, and one speaker mounted in the base of the pendulum and controlled by the Arduino. The Arduino itself is controlled by custom software that is constantly sending and receiving data to make sure that the pendulums are coordinated to coincide with the score that is playing.

This project shows that you can use Arduino to great effect when combined with an understanding of other disciplines, such as art, mechanical engineering, and architecture. On their own, each of the elements of this project is relatively simple: controlling a motor, controlling an LED, and playing a melody. The challenge is when the scale of each of these increases. Controlling high-powered motors requires knowledge of loads and mechanics; controlling lots of LEDs requires an understanding of controlling higher voltage and current; and playing high-quality sound requires specific hardware.

Chapter 12 introduces you to shields that can make a lot of functions, similar to those used in Chorus, a lot easier to achieve with your Arduino.

Further reading

You can find the project page at UVA's website: `www.uva.co.uk/works/chorus`. In addition, an excellent paper by Vince Dziekan looks at both UVA's working practice and the Chorus project in detail: `http://fibreculturejournal.org/wp-content/pdfs/FCJ-122Vince Dziekan.pdf`.

Push Snowboarding

Push Snowboarding is a collaborative project between Nokia and Burton that aimed to visualize all the data about your snowboarding run. Vitamins Design Ltd (now Special Projects) was tasked to design and build a set of wireless sensors to communicate with a mobile phone in the snowboarder's pocket. Vitamins was founded by Duncan Fitzsimons, Clara Gaggero, and Adrian Westaway, who dub themselves a "design and invention studio." They are located in London, where they develop and make products, experiences, concepts, and systems for their clients. They have worked on a variety of projects, which is best described in their own words, having "worked on fashion catwalks and in operating theatres . . . with pro snowboarders and factory workers . . . for experimental theatre as well as techno-phobic pensioners."

For the snowboarding project, Vitamins designed a set of 3D printed sensor boxes (shown in Figure 8-3) to measure galvanic skin response, heart rate, balance, 3D motion, position, geographical location, speed, and altitude. This data was then layered over a live video of the snowboarding run to show the links between different situations and the physical response of the snowboarder. This project is an excellent example of making well-finished Arduino products that are closer to customized products than prototypes.

How it works

Each sensor is in a waterproof enclosure with a power button. When switched on, the Arduino talks wirelessly to the smartphone, communicating whatever data it finds. From there, the smartphone, with its superior processing power, crunches the numbers and compiles the data to present the snowboard run visually to the snowboarder.

The first challenge of this project was the size. Many sensors can be tiny, but the Arduino itself creates a big footprint, and if a snowboarder were wearing several sensor boxes, the data would be useless if the boxes were obstructing his or her movement. For this reason, each sensor box has an Arduino Pro Mini, which is wafer thin and only 18mm x 33mm. The same is true for the power source, which is provided by a rechargeable lithium battery, the same kind found in model aircraft and smartphones.

FIGURE 8-3: Snowboarding sensors in customized sensor boxes.

Courtesy of Vitamins Design Ltd

The project uses a variety of sensors: pressure sensors on the feet to judge balance; inertial measurement units (IMU), also referred to as degrees of freedom sensors, which are used to find the 3D orientation of the snowboarder; galvanic skin-response sensors to monitor sweat levels of the snowboarder; and a heart rate monitor to track the snowboarder's BPM.

This data is sent back to the phone wirelessly using a Bluetooth module located in each sensor box. The Bluetooth module is small, has a secure connection, and will reliably work over the short distance between the sensor box and the phone in the snowboarder's pocket. The data can then be combined with other data gathered by the phone, such as Global Positioning System (GPS), and then formatted by custom software on the phone.

Each of the sensors in this project is available from most online Arduino-related shops, along with examples and tutorials of how you can integrate them into your own project. The execution of this project is what makes it a great example for aspiring Arduinists. The combination of sensors provides snowboarders with a wealth of information that could be used to improve their technique or rate their performance in different ways. In addition, these products were made to survive in tough environments. The electronics, are located behind a tough case to protect them from impact and are carefully padded to protect them from shock, but they are also accessible. The case even includes a damp course to trap any moisture that may get in.

Further reading

You can find more details about the project on the Vitamins website at `specialprojects.studio/project/push-snowboarding/`.

Baker Tweet

Baker Tweet is a project by Poke that tweets the latest freshly baked goods from the Albion Café as they come out of the oven. Poke is a creative company based in London and New York that specializes in all things digital. The project resulted from an assignment Poke London was given to put the newly opened Albion Café on the map. It proposed a "magical box" that allowed the café to announce its freshly baked goods using Twitter so that the locals would know when to visit to get the freshest bread, croissants, and buns. You can see a picture of the Baker Tweet device in Figure 8-4, or pop into the café if you're in the neighborhood!

FIGURE 8-4: Baker Tweet in the Albion Bakery.

Courtesy of Poke London

How it works

The "magical box" that produces Twitter messages has a simple interface consisting of a dial, a button, and an LCD screen. The café's employee uses the dial to select the freshly baked item to be tweeted, the tweet is displayed on the LCD screen, the person pushes the button to send the tweet, and feedback on the LCD appears when the tweet is successfully sent. This simple interface is ideal for a busy café, requiring little time to update the online status. To add new items, the project uses an easy-to-use web interface that updates the list.

At the heart of this project is an Arduino that communicates with the Internet. This can be accomplished either with a wired Ethernet connection, using the Arduino Ethernet shield, or wirelessly, using a Linksys Wi-Fi adapter. The dial provides an analog input; the button provides a digital input. Combined with the LCD screen, they create a user interface for easily moving through the list and sending tweets. The prototype is encased in a sturdy enclosure to prevent dough and floury hands from damaging the circuit.

This project shows a great application for Arduino that performs the often time-consuming activity of updating an online status quickly and easily. The prototype is also built to suit its environment, being robust enough with its stainless steel inputs to survive constant use. It's also easy to clean considering the messy bakery environment. The complexity of this project lies in the communication with the Internet, which for the more web savvy of you may be an area of interest, allowing your physical project to send data to the World Wide Web and vice versa.

Further reading

You can find much more information on the Baker Tweet site at `http://www.bakertweet.com`) and the Poke London projects page at `pokelondon.com/work/baker-tweet/`. Also, take a look at a few excellent prototyping photos that show the development of the project from breadboard to bakery (`flickr.com/photos/aszolty/sets/72157614293377430/`).

The National Maritime Museum's Compass Lounge and Compass Card

The Compass Lounge was developed as part of a new wing of the National Maritime Museum in London. London-based design studio Kin designed and developed the interactive areas of the Compass Lounge, which used a number of

Arduino projects behind the scenes to allow the public to interact with the digital archive of the museum as well as the physical pieces on display.

A set of digital plan chests (*plan chests* are large drawers that usually store large prints, blueprints, or paperwork) allow visitors to browse the museum's online archive and access the most popular items in high resolution. When the plan chest is opened, a large touchscreen is activated, and the visitor can browse the items on display. While the visitors are browsing, a hidden LED display lights up through the wallpaper (as shown in Figure 8-5), displaying the reference number of the current item so that it can be found in the physical archive.

FIGURE 8-5: The hidden LED display lights up underneath the wallpaper.

Courtesy of Kin

Another aspect of the Arduino-powered part of this project is the Compass card system. Each visitor is given a card to use to collect items from all over the museum. Next to certain items are collection points where the card is scanned to collect the item digitally and stamped to leave a physical mark on the card to show the visitors journey through the museum. Visitors can browse their collected items in the Compass Lounge or at home in a browser.

How it works

The plan chests use a couple of simple Arduino projects to complement the digital content displayed on the large touchscreens.

The Arduino is used to activate the screens when they're opened. If the screens are left on throughout the day when not in use, the image can often be burned into the screen, leaving shadows when the content is changed. Setting the background to black when the plan chest is closed reduces this screen burn and extends the lifetime of the monitors. This result is accomplished by a microswitch on the back of each drawer — no different from the setup for the Button sketch in Chapter 6. Whenever the button is pressed, a character is sent over the serial port to tell the monitor to go black. This sort of communication is covered in Chapter 7.

Above the plan chests is a hidden LED display made up of rows of LEDs aligned in a grid. The letters are sent over the serial port as a string from the custom software that displays the images so that the correct number appears with the correct image. Because the LEDs are so bright, they can shine through the fabric wallpaper when on and remain hidden when not. This is an excellent example of using a premade product in a different arrangement to suit the purpose.

The Compass cards are also a great example of using (relatively) old and existing technologies in a new and interesting way. The Compass cards use barcodes to know which card is being scanned. This returns a number to the Arduino that can be forwarded to a central server that coordinates the barcode number and the scanner number to identify where the card has been scanned and by who.

All this information is sent over the Ethernet to a server using an Ethernet shield, where the information can be collated and outputted as needed. Arduinos are doing the relatively complicated task of relaying data over a network without the need for computers at every collection point. This scenario not only cuts the cost of data transfer but also makes the operation of the network easier because an Arduino requires no startup or shutdown procedures. This digital system works alongside a physical stamping system to mark the cards with an embosser, which makes an impression into the card when the lever is pushed. (The inside of the Compass card collection point is shown in Figure 8-6.)

This collection of projects illustrates how many applications you can bring together to create an experience, providing many different forms of interaction and feedback. It's also a great example of an extreme-use case for Arduino. Many projects, prototypes, or installations show that Arduino can work, but it is often regarded as unreliable and seen as a temporary solution rather than a long-term one. In this case, the museum demanded a reliable and robust solution, and Arduino was more than capable of the task when used correctly.

FIGURE 8-6: A look inside the Compass card collection points.

Courtesy of Kin

Further reading

You can find much more information as well as illustrations on the Kin project page at `kin-design.com/commissioned-work/arts_culture/compass-card-system-national-maritime-museum/`.

The Good Night Lamp

The Good Night Lamp project is an Internet-connected family of lamps founded by Alexandra Dechamps-Sonsino. Each family is made up of a Big Lamp and numerous Little Lamps. When the Big Lamp is on, the connected Little Lamps turn on as well, wherever they are in the world. They allow loved ones to remain in touch with each other by simply going about their daily routine, without actively using any application. The Good Night Lamp is currently in development using Arduino as the basis for the prototypes. A set of lamps appears in Figure 8-7.

How it works

The system for the preproduction prototypes of the Good Night Lamp is relatively simple. The Big Lamp is a functional light operated with a pushbutton, similar to the ButtonChangeState example in Chapter 10. It illuminates the light and sends the ID number of the lamp and state of the light to a web server by using an Arduino Wi-Fi shield.

FIGURE 8-7: Whenever the Big Lamp is turned on, the Little Lamp turns on as well, wherever it is in the world.

Courtesy of Good Night Lamp Ltd

Somewhere else, maybe on the other side of the world, the Little Lamps download the state of the Big Lamp by using the same Wi-Fi shield. If a Big Lamp is on, any Little Lamps that are paired with it are also turned on.

The lamps themselves are high-power LED bulbs running off 12V and requiring 0.15 amps for the Little Lamps and 0.4 amps for the Big Lamp. These LED bulbs are ideal for the high brightness needed for functional lamps and require a transistor circuit to run them, the same as the one explained in Chapter 7.

This project is a great example of using Arduino to produce a prototype of a product with relatively complex behavior. Using Arduino and other tools enable you to develop the electronics inside a product to reliably demonstrate its behavior.

Further reading

If you would like to read more about the Good Night Lamp, go to the product home page at `goodnightlamp.com`.

Little Printer

Little Printer (shown in Figure 8-8) is a miniature home printer developed by Berg, a design consultancy based in London. Using your smartphone, you can manage content from the web to print all sorts of information to create your own personal newspaper. In between your phone and the web is BERG Cloud, which does all the heavy lifting. Little Printer is built with custom hardware and software but was prototyped using Arduino.

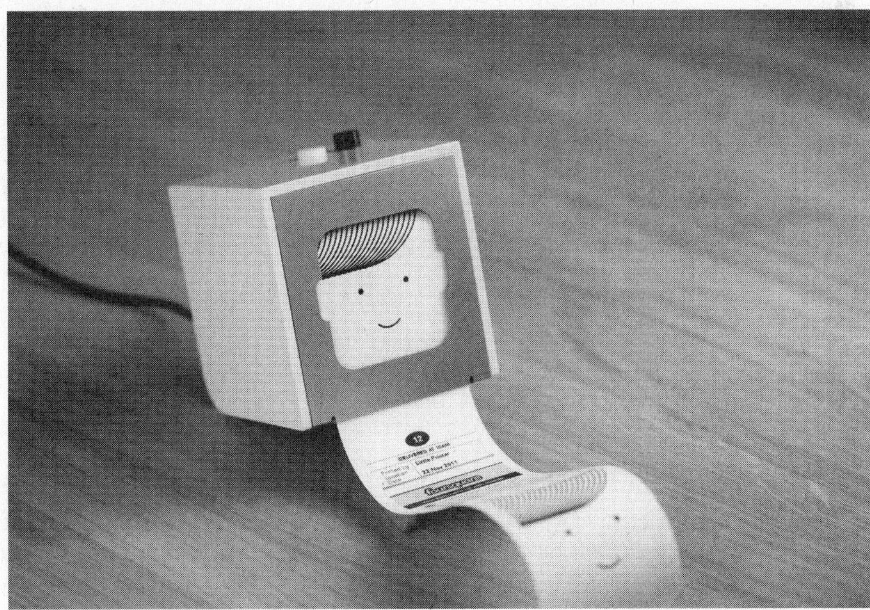

FIGURE 8-8: A Little Printer ready to print whatever data you'd like.

Courtesy of Berg

How it works

The Little Printer is made up of several parts, starting with the printer itself, which is a thermal printer, similar to those used to print shopping receipts. Thermal printers communicate over serial, so with the right code, it's possible to talk directly to a device like an Arduino. Adafruit stocks one such printer that is easy to integrate into your own Arduino projects (go to www.adafruit.com/products/597).

The printer itself is powered but talks wirelessly to the BERG Cloud Bridge, a small device that handles all data in a way similar to your home router. Data is then sent and received using a wired Internet connection, the same as that used on the Arduino Ethernet shield.

In early prototypes, XBee wireless modules — the same as those used on the Arduino wireless shield — handled communication between the Little Printer and the BERG Cloud Bridge.

Much of the complexity of this product is handled in the BERG Cloud, where data is gathered, sorted, and then sent to the printer as requested.

The Little Printer is a great example of a product that has used Arduino to develop and refine the idea, before being developed further into a product with its own custom hardware. It also shows that it is possible for your Arduino project to make use of the abundance of data on the Internet.

Further reading

Sadly, Little Printer is no more, but it remains a textbook example of what's possible with Arduino. To find out more about their journey, head over to littleprinterblog.tumblr.com.

Flap to Freedom

Flap to Freedom was a game by ICO, a London-based design consultancy created as part of the V&A Village Fete, described by the V&A as a "contemporary take on the British Summer Fete." It took the form of a game where two players would race, head-to-head, to help a chicken to escape from its battery farm (as shown in Figure 8-9). If a player flaps his arms, the chicken does also, but if he flaps too quickly, the chicken tires out. Each race was timed and put on the "pecking order" leader board to determine the winner.

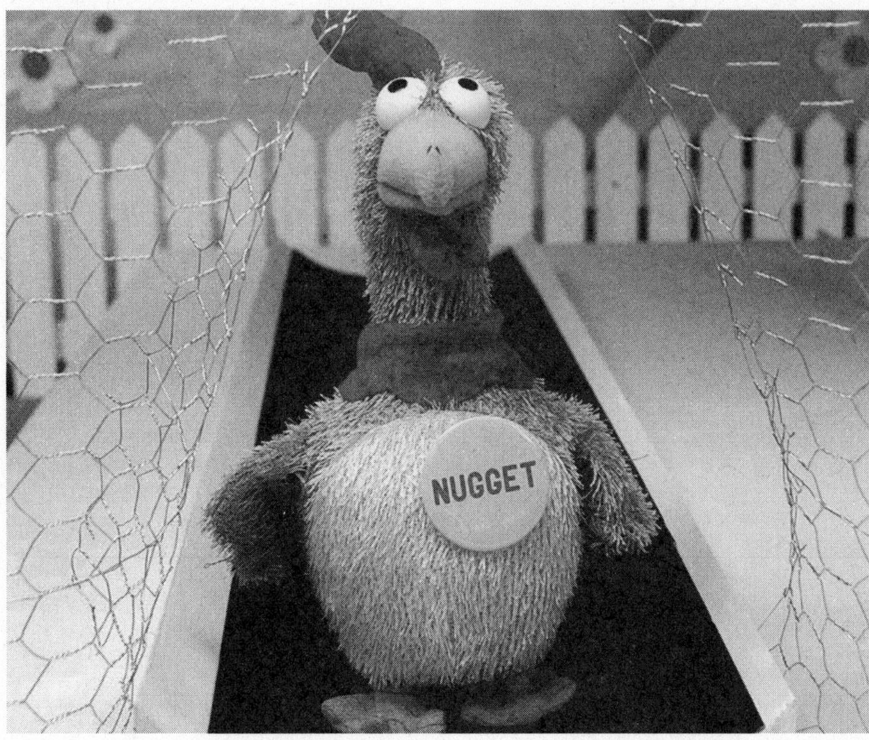

FIGURE 8-9: Chicken run!

Reproduced by permission of ICO Design Printers Ltd

How it works

For the Arduino-related project, the existing toy chickens used in this race were taken apart and given new custom circuits that communicated wirelessly with software on a computer. The circuits inside the chickens were remade to give an Arduino control of the various motors that moved the wings, beak, and feet. These were each told what to do by using wireless modules, similar to the XBee modules on the Arduino wireless shield, which communicated with a piece of custom software on a computer.

The software used was openFrameworks. Using a hidden webcam, the program analyzed people's movements to determine how fast the chickens should waddle and then sent the appropriate signal to each chicken.

This fun and engaging application for Arduino allows people of all ages to play a game and have instant physical feedback from the toy chickens. The program also illustrates how to hack an existing toy to give it a different use and can be applied to other products.

Further reading

You can find much more to read on the ICO project page at `icoex.co.uk/work/project/flap-for-freedom`. Benjamin Tomlinson and Chris O'Shea worked on the technical side of the project; for a more technical description, go to Chris O'Shea's project page at `www.chrisoshea.org/flap-to-freedom`.

IN THIS CHAPTER

» Learning all about soldering

» Getting all the right kits for the job

» Assembling a shield

» Moving from the breadboard to strip board

» Preparing your project for the real world

Chapter 9
Soldering On

In previous chapters, I cover in great detail how to assemble circuits on a breadboard. If you read those chapters, you most likely already have a few ideas that build on or combine a few of the basic examples, so you may be asking, "What do I do next?"

This chapter takes you through the process, or art, of soldering. You discover all the tools you need to get your project ready for the real world. No more precariously balanced breadboards or flailing wires. From this point on, you'll know what you need to solder circuit boards that last.

Understanding Soldering

Soldering is a technique for joining metals. By melting metal with a much lower melting point than the metal you're joining, you can link pieces of metal to form your circuit. Mechanical joints are great for prototyping, allowing you to change your mind and quickly change your circuit, but after you're sure of what you're making, it's time to commit.

You use a soldering iron or solder gun to melt *solder*, a metal alloy (mixture of metals) with a low melting point, and apply it to the joint. When the solder has cooled around the pieces that are being connected, it forms a secure chemical bond rather than a mechanical bond. This method is a far superior way to fix components in place, and bonded areas can still be melted and resoldered, if needed.

But why do you need to mess with soldering at all? Picture this: You have your circuit on a breadboard and you're ready to use it, but every time you do, the wires fall out. You could persevere and keep replacing the wires, but you risk replacing the wrong wire and damaging the Arduino or yourself. The best solution is to make a soldered circuit board that's robust and can survive in the real world.

A solderless breadboard allows you to quickly and easily build and change your circuit, but after you know that it works, you need to solder the circuit to keep it intact.

Creating your own circuit board is also an opportunity to refine your circuit by making circuit boards that fit the components. After you know what you want to do, the process of miniaturization can start and you're eventually left with a circuit board that takes up only the required space and no more.

Gathering What You Need for Soldering

Before you dive in to soldering, make sure that you have what you need to get the job done. Read on to find out more.

Creating a workspace

For your soldering adventures, what you need above all is a good workspace. Having a good workspace can make all the difference between a successful project and hours spent on your hands and knees, swearing at cracks in the floorboards. A large desk or workbench would be perfect, but even the kitchen table, if clear, will work. Because you're dealing with hot soldering irons and molten metal, it's a good idea to cover the surface with something you don't mind damaging. A cutting mat, piece of wood, or piece of cardboard will do fine for this purpose.

Your workspace should be well lit as well. Make sure that you have ample daylight by day and a good work light at night to help find those tiny components.

It's also good to have easy access to a power source. If your soldering iron functions at a fixed temperature and with a short lead connected directly to a plug, it can be especially important to have a plug nearby. If you overstretch your lead, you run the risk of pulling the iron off the table and burning anything it touches. A tabletop power strip or multi-plug is the best solution because it provides power for your laptop, your lamp, and your soldering iron.

A comfortable chair is always important. Also remember to stand up every half hour or so to prevent back cramp. You can easily get drawn into soldering and forget what a horrible posture you're in.

Solder fumes, although not lethal, are not good for your lungs, so make every attempt to avoid breathing them. Always work in a well-ventilated area. It's also advisable to work with lead-free solder, as mentioned later in this section.

TIP

If you're working at home and are under pressure from other people in your house to not cover every surface with bits of metal, you should designate a soldering surface. This surface could be a rigid, wooden surface that can fit all your kit and can be moved, neatly packed away, or covered when not in use. This arrangement saves you the chore of unboxing and packing up every time you want to solder — and keeps everyone else happy as well.

Choosing a soldering iron

The most important tool for soldering is, obviously, a soldering iron or solder station. You have a huge variety to choose from, but they're generally divided into four types: fixed temperature, portable, and temperature-controlled soldering irons, and complete solder stations. I describe each type in the following sections and provide a rough price from my local retailers. When you have an idea of what you want, shop around locally to see what deals you can find. If you're lucky, you may even find some high-quality second-hand gear on eBay!

Fixed-temperature soldering iron

A *fixed-temperature soldering iron* (see an example in Figure 9-1) is normally sold as just an iron on a piece of electrical cable with a plug at the other end. They are usually sold with a sponge and a flimsy piece of bent metal as a stand. Others that are slightly better have a plastic stand with a place to put your sponge and a decent spring-like holster for your iron.

Fixed-temperature irons are adequate but offer no control over the temperature of your iron. They are sold with a power rating or wattage, which is of little help to most people. On these fixed-temperature irons, a higher wattage means a higher temperature, although that can vary wildly from manufacturer to manufacturer.

WARNING

This variation can cause trouble with more delicate components because a high temperature is quickly conducted and melts integrated circuits and plastic components.

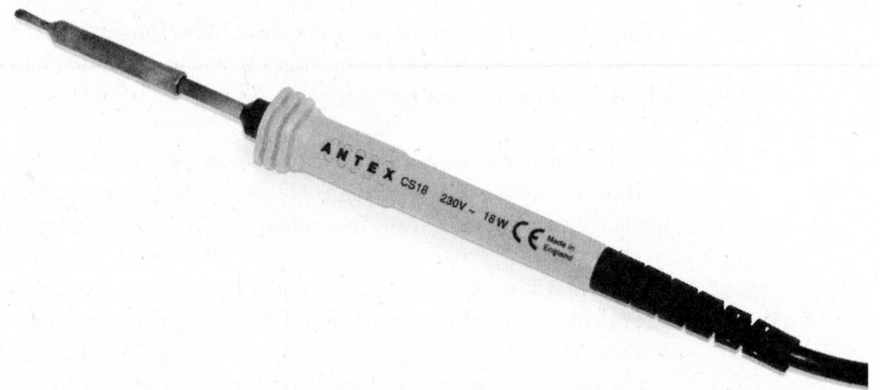

FIGURE 9-1: A basic fixed-temperature soldering iron.

A quick study of RadioShack shows soldering irons in a range of powers, from 15W to 60W, which could cover a temperature range of approximately 400° F to 750° F. The difficulty is in finding an iron that is hot enough so that it heats the part that you want to heat quickly and allows you to melt the solder before the heat spreads. For this reason, a low-temperature iron can often do more damage than a higher temperature one. If you go too high, you encounter other problems, such as having the tips erode faster and running a higher risk of overheating parts.

If you want to go with a fixed-temperature iron, my advice is to start with a midrange iron. I recommend a 25W iron as a good starter, which costs in the region of $22 from RadioShack.

Portable soldering iron

A *portable soldering iron* does away with cables and electricity in favor of gas power. It burns butane gas (more commonly known as lighter fuel) to produce its heat. A portable iron has a wattage rating to allow comparison with other irons, but unlike the fixed-temperature ones described in the preceding section, this rating indicates the maximum temperature, which can be lowered by use of a valve.

WARNING

The flames burn around the edge of the iron, which can make them awkward to use on precise joints, so I recommend using a portable iron only if necessary.

A portable iron (shown in Figure 9-2) is great for tight situations when an extension lead just won't reach, but it's a bit too expensive and wasteful to be used often. It's also considered a lot more dangerous than conventional irons by most airport security because of the use of gas. So if you're planning on doing any soldering abroad, take an electric iron.

Butane soldering irons vary in price, but you can usually find them in the range of $45 to $90. You also need to buy additional butane refill cans for $7.50 each.

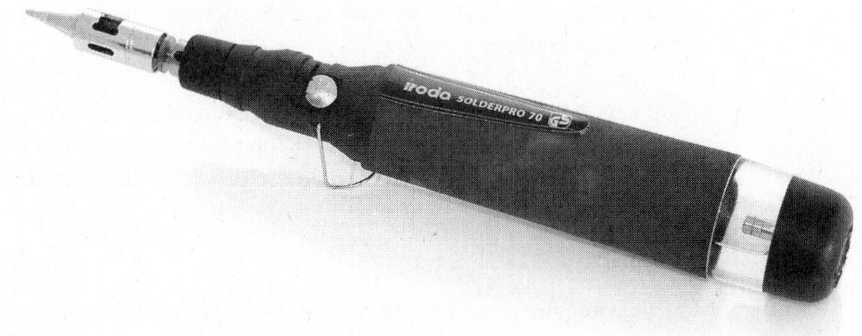

FIGURE 9-2: A butane soldering iron. Watch those eyebrows!

Temperature-controlled soldering iron

A *temperature-controlled soldering iron,* shown in Figure 9-3, is preferable to the fixed-temperature variety because it gives you more control at a reasonable price. This increase in control can make all the difference between melting and burning. Temperature-controlled irons should have a wattage rating, but it will be the maximum power possible. For this type of iron, a higher wattage is preferable because it should give you a better range of higher temperatures.

FIGURE 9-3: A temperature-controlled soldering iron.

A temperature control dial allows you to scale your temperature range up or down as needed. The difference between this type of temperature control and the control when using more accurate solder stations is that you don't have a reading of the current temperature. Most temperature-controlled irons have a color wheel indicating warm to hot, so a bit of trial and error may be necessary to get just the right temperature.

You can get an affordable temperature-controlled soldering iron for around $30. Because you have more control over the temperature of the iron, you have greater control and more longevity than you have with a fixed-temperature iron.

Solder stations

After you gain some experience (and can justify the expense), a solder station is what you'll want. A *solder station* is usually made up of an iron, a stand or cage for the iron, a temperature display, a dial or buttons for temperature adjustment, and a tray for a sponge. It can also include various other accessories for de-soldering or reworking, such as a hot air gun or vacuum pickup tool, but these are generally for more professional applications and not immediately necessary.

You can find many available brands of solder stations, but one of the most reputable and widely used is Weller (see Figure 9-4). Of that company's stations, I recommend the WES51 (120V AC), the WESD51 (120V AC), or the WESD51D (240V AC). My WES51 is still performing excellently after four years of use. Note that the 120V irons need a transformer to step the voltage down in countries that use 240V AC; the transformer can often be heavier than the iron itself!

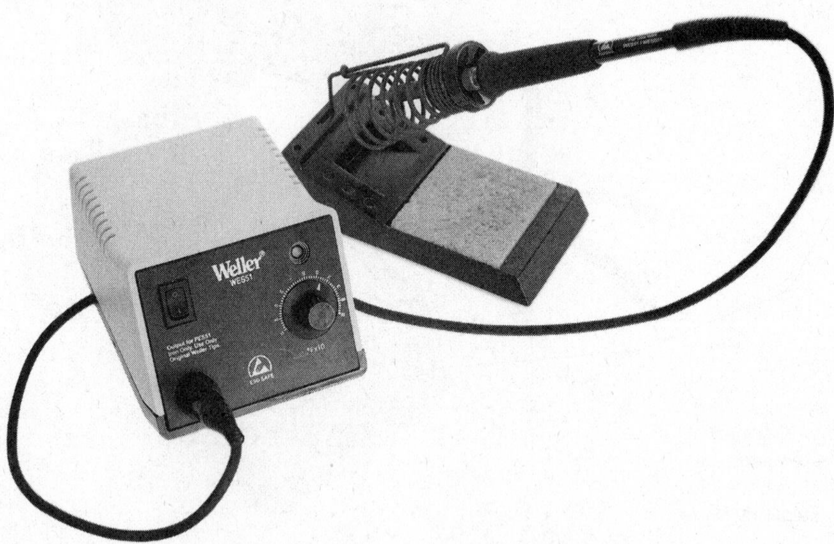

FIGURE 9-4: A Weller solder station.

Before using a Weller solder station, I owned a cheaper temperature-controlled soldering iron that did very well for a number of years. As I cover in more detail later in this chapter, the best way to maintain and get the most out of your iron is to use it correctly. Leave a little melted solder on the tip to protect it when it's not in use. Then, before using the iron again, use the sponge to remove the solder.

Soldering iron tips

Regardless of which iron you buy, I also recommend buying a few spare tips because they eventually degrade. You'll find a variety of tips for different purposes, too, so it's good to have a selection to cover different needs.

Solder

Solder is what you use to join your circuits. Although many different types of solder are used with different combinations of metals, they commonly fall into two categories: lead and lead free. Many people prefer lead solder, finding it easier to work with, perhaps because it has a lower melting point, so your iron can be cooler and less likely to damage components.

Lead poisoning has been known about for a long time, and attitudes about using it have begun to change. Lead pipes were switched to copper as recently as the 1980s, and the use of lead solder in consumer electronics was addressed in 1996, when the Restriction of Hazardous Substances Directive (RoHS) and the European Union Waste Electrical and Electronic Equipment Directive (WEEE) addressed the use and disposal of certain materials in electronics.

You commonly find RoHS on components that comply with their guidelines, which should, therefore, be better for the environment when disposed of. Commercially, US and European companies enjoy tax benefits for using lead-free solder (shown in Figure 9-5), but lead is still widely used in the rest of the world. Steven Tai, a colleague of mine, visited China to complete a project we were working on. When he asked where he could buy lead-free solder, he was laughed at outright because lead-free solder was not only was unheard of in most cases but not even available! For the more conscientious Arduinists, most electronics supplier and shops offer lead-free solder that contains other metals, such as tin, copper, silver, and zinc. From my experience, lead-free solder works just fine for any Arduino projects, so if you want to do your bit for the environment and avoid using lead in your work, please do!

Another variety of solder is flux-cored solder. *Flux* is used to reduce oxidization, the reaction on the surface of a metal when it reacts with the air (oxygen), as with rust on an iron anchor. Reducing the oxidization allows a better, more reliable connection on your solder joint and allows the solder to flow more easily and fill the joint. Some solders have flux in the core of the solder, dispensing flux to the joint as the solder is melted. You sometimes see smoke as you melt your solder; in most cases, the smoke is from the flux burning off. You can be sure that you have flux core if, when you cut the solder, you see a black center surrounded by a tube of solder.

FIGURE 9-5:
Lead-free solder.

Always work in a well-ventilated area and avoid breathing the solder fumes no matter which solder you're using. Solder fumes are not good for you, and neither is eating solder. Always wash your hands and face thoroughly after soldering. Sometimes the flux in solder can spit, so wear clothes that are not precious — and definitely use eye protection.

Third hand (helping hand)

Sometimes you just don't have enough hands to hold the fiddly electronics that you're trying to solder. It would be great to have someone nearby with asbestos hands to hold the tiny red-hot pieces of metal, but failing that, you can use a little device known as a *third hand* or a *helping hand,* as shown in Figure 9-6. A third hand is a set of crocodile clips on an adjustable arm. You can arrange the clips to help you get your component, circuit board, and solder into place.

A third hand costs from $5 to $50 and can be extremely useful for holding circuit board at an angle or holding components together while you work on the soldering. The drawback is that setting it up can be tricky. If you're doing lots of solder joints, you may spend a lot of time loosening, adjusting, and retightening. If you do purchase a third hand, make sure that the parts are all metal. Plastic parts, such as the grips on the vices, will not stand up to much use.

Adhesive putty

A good alternative to a third hand is adhesive putty, such as Bostik's Blu-Tack, UHU's White Tack, or Locktite's Fun-Tak. Rather than use a mechanical device to grip a component or a circuit board, you use the adhesive putty to hold the objects you're soldering in place on one side of the board, leaving you free to work on the other side of the board without the components or circuit board moving. You can also use the adhesive putty to tack your work to your work surface, stopping it from moving around the surface as you solder. After the solder joints are done, you remove the adhesive putty, which you can reuse. Note that putty goes extremely soft if it is heated and takes a while to return to its usual tackiness. After it cools, you can roll the ball of putty along the circuit board to remove any remaining bits.

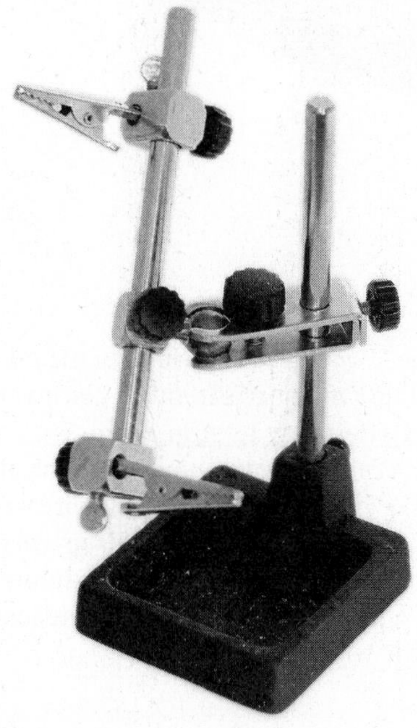

FIGURE 9-6: Give the man a hand!

Wire cutters

A good pair of wire cutters or snips is invaluable. Many pairs of wire cutters have a rounded, claw-like shape. These are tough but can be difficult to use when cutting in confined spaces or pinpointing a specific wire. Precision wire cutters have a more precise, pointed shape that is far more useful for the vast majority of electronics work.

Note that wire cutters are good for soft metal such as copper but do not stand up to tougher metals such as paper clips or staples. Figure 9-7 shows a pair of pointed wire cutters.

FIGURE 9-7: Pointed wire cutters are good for getting into and out of tight spots.

Wire strippers

To connect wires to your project, you need to strip back the plastic insulation. You can do this stripping with a knife if you're careful or don't value your fingertips, but the quickest, easiest, and safest way to strip a wire is to use a wire stripper. The two kinds of wire strippers are manual and mechanical (see Figure 9-8). Manual wire strippers are like clippers but have semicircular notches made to various diameters. When the wire stripper is closed on wire, it cuts just deep enough to cut through the plastic sheath but stops before it hits the wire. Mechanical wire strippers work with a trigger action to remove the insulation on the wire without any need to pull on the wire.

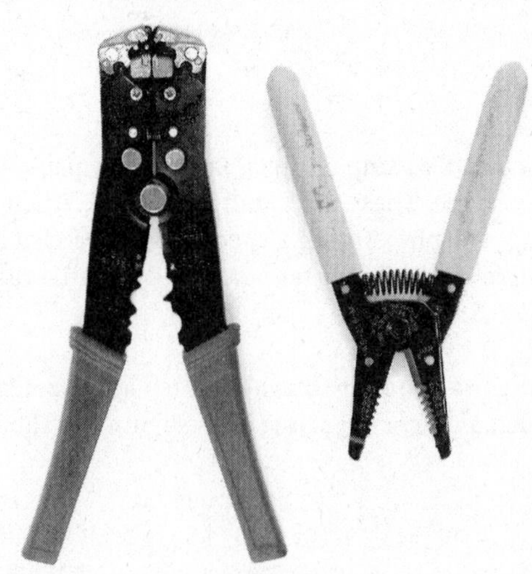

FIGURE 9-8: Manual and mechanical wire strippers.

Mechanical wire stripers are a great timesaver but can be less reliable in the long run because the mechanisms are more likely to fail than are those in the simple manual ones.

Needle-nosed pliers

Needle-nose pliers, as with a solderless breadboard, are a great help for getting to those hard-to-reach places. They're especially useful when soldering because they spare your fingers from any excess heat. I cover needle-nosed pliers in more detail in Chapter 4.

Multimeter

A multimeter is a great help for testing your circuits. When you're soldering connections, the continuity-testing function can be a great help for verifying that the solder joints are as they should be and are connected to the right places. See Chapter 4 for more about working with a multimeter.

REMEMBER

When testing the continuity, always unplug any power supplies connected to your circuit to avoid false bleeps.

Solder sucker

Everyone makes mistakes, and they can be more difficult to undo when you're dealing with hot metal. One tool to keep handy for fixing mistakes is a solder sucker, shown in Figure 9-9, or a de-soldering gun. Each of these tools is a pneumatic pump that sucks molten solder straight off the surface of a circuit board. Most have a piston that can be pressed down and will lock when all the air is pressed out. When you hit the trigger, a spring pushes the piston back out and sucks any molten solder into the piston chamber. Next time you press the piston down, it pushes out any solder that was removed. Using this type of tool takes a bit of practice because you need to heat the solder with your soldering iron in one hand and suck it away with the sucker in the other.

Solder wick

Another method of removing solder is to use solder wick (see Figure 9-10), also known as copper braid or de-soldering wire. Solder wick is a copper wire that has been braided, and you buy it in reels. It provides lots of surface area for solder to grip into, to remove it from other surfaces. Place the braid on the joint or hole that has too much solder, and hold your soldering iron on top of it to heat the solder wick. Apply some pressure, and the solder continues to melt and fill the gaps in between the braids of the copper wire. Remove the wick and the soldering iron together, and the solder should be cleared. If it's not, repeat as necessary. After the solder is cleared, you can cut off the used solder wick and dispose of it.

FIGURE 9-9: A solder sucker can save your project.

FIGURE 9-10: Solder wick is great for those more stubborn mistakes.

WARNING

Do not pull the solder wick away if the solder has cooled. You risk pulling off the metal pads on your board, making it unusable, because the solder wick is attached to the board.

TIP

Solder suckers and solder wick are both equally effective at removing solder, but they are each suited to different situations and require an element of dexterity and skill to use. If you're worried about overheating components, a solder sucker is more suitable; if you can't remove all the solder with the sucker or can't get close enough to the components with it, the solder wick may be a better option. I advise getting both to prepare yourself for any situation.

Equipment wire

Equipment wire is the general name given to electronics wire. It's the same as the jump wires you may have in your kit but is unfinished. You buy it in reels. Equipment wire can be either single-core or multicore. Single-core wire is made up of one solid piece of wire and is malleable, so it holds its shape if bent, but it snaps if it is bent too much. Multicore wire is made up of many fine wires and can withstand a great deal more flexing than single core, but it does not keep its shape if bent. To use equipment wire, you need to cut it to length and strip the insulation off the ends of to reveal the wire underneath.

Wire also comes in different diameters, indicated with numbers such as 7/0.2 or 1/0.6. In this format, the first digit is the number of wires in the bundle and the second is the diameter of those individual wires. So 7/0.2 is a bundle of 7 wires, each measuring 0.2 mm, making it multicore; 1/0.6 is one single-core wire with a diameter of 0.6mm.

When you're starting out, it can be difficult to know what type of wire to invest in. Wire is cheaper when bought by the reel, but you don't want to be stuck with wire that isn't fit for your purposes. As a general guideline, I have found that multicore wire is the most versatile and robust for most applications. The 7/0.2 diameter should fit most PCB holes. I also recommend having three colors — red, black, and a color to signify your signal wires. With three reels of this type and size, you should be able to complete most projects. Some hobby electronic shops also supply lengths in various colors, such as the ones by Oomlout shown in Figure 9-11.

FIGURE 9-11: Equipment wire in various colors.

Staying Safe while Soldering

With a few simple precautions, you can solder safely. Remember that soldering is not dangerous if you take proper care — but it can be if you don't. Please keep the tips in this section in mind whenever you're soldering.

Handling your soldering iron

A soldering iron is safe if used correctly but is still potentially dangerous. The iron has two ends, the hot end and the handle. Don't hold it by the hot end! The correct way to hold a soldering iron is like a pen, between your thumb and index finder, resting on your middle finger. When you're not using the iron, keep it in its holster or cage, which helps to dissipate heat and prevent accidental burns.

Keeping your eyes protected

You must wear proper eye protection when soldering. Solder, especially the flux-cored kind, has a tendency to spit when heated. In addition, when you use clippers to neaten your circuit board, the small bits of metal you cut off often shoot around the room if they're not held down. Also, if you're working in a group, you need to protect yourself from the person next to you. Safety goggles are relatively inexpensive depending on the amount of comfort you want, and they're a lot cheaper than eye surgery.

Working in a ventilated environment

Breathing in fumes of any kind is bad for you, so it's important to always solder in a well-ventilated environment. Also make sure that you're not working under any smoke alarms because the fumes from soldering can set them off.

Cleaning your iron

Your soldering iron should come with a sponge, which you use to wipe away excess solder. You should dampen the sponge but not use it soaking wet, so make sure to squeeze out excess water. When you're heating solder, it oxidizes on the tip of the iron. If the tip itself oxidizes, it can degrade over time. To prevent oxidizing the tip, leave a blob of solder on the end of the iron while it's in its cage. Doing so makes the blob of solder oxidize rather than the tip, and you just wipe off it off using the sponge the next time you need the iron.

Don't eat the solder!

Although the chemicals and metals in solder are not deadly, they are definitely not healthy and can cause irritation. While soldering, avoid touching your face and getting solder around your eyes and mouth. It's also a good idea to wash your hands (and face if needed) after soldering.

Assembling a Shield

Soldering requires learning a great amount of technique, and you develop good technique with practice. In this example, you find out how to assemble an Arduino shield. A *shield* is a specific printed circuit board (PCB) that sits on top of the Arduino to give it a function. (You learn more about shields in Chapter 12.)

There are different shields for different functions. The one used in the example is the proto shield kit (shown assembled in Figure 9-12), which is essentially a blank canvas to solder your project onto, after prototyping it on a breadboard. In this example, you see how to assemble the bare minimum of the kit to attach it to your Arduino and then how to build a simple circuit on it.

FIGURE 9-12: A complete proto shield.

As with many Arduino kits, you need to assemble this shield yourself. The basic principles of soldering remain the same but may vary in difficulty as you encounter smaller or more sensitive components.

USING STRIPBOARD RATHER THAN A PCB

Specially designed shields are made to fit your Arduino perfectly but can often be relatively expensive. Stripboard, or perfboard as it's sometimes known, provides a cheap and highly versatile alternative. *Stripboard* is a circuit board with strips of copper and a grid of perforated holes that you can use to lay out your circuit in a similar way as on a breadboard. An example of stripboard appears in the following figure.

The pitch of the holes and the layout of the copper strips can vary. The most useful pitch for Arduino-related applications is the same as the pitch on the Arduino pins, 0.1 inches (2.54mm), because that pitch allows you to build on the layout of your Arduino to make your own custom shields. You can buy stripboard in various arrangements of copper strip as well, commonly either long copper columns that run the length of the board or sets of columns three rows deep (usually called tri-board).

Laying out all the pieces of the circuit

When assembling a circuit, your first step should always be to lay out all the pieces to check that you have everything you need. Your work surface should be clear and have a solid-colored cover to make things easy to find.

Figure 9-13 shows the Arduino proto kit laid out in an orderly fashion. It contains the following:

- Header pins (40x1)
- Header sockets (3x2)
- Pushbuttons
- LEDs (various)
- Resistors (various)

FIGURE 9-13: All the parts of the shield laid out.

Some kits may ship the PCB only and leave you to choose the headers that are connected. Remember that there is no right or wrong way as long as the assembly suits your purpose.

To assemble this shield, you can work from a picture to see the layout of the components, but for more difficult ones, you usually have instructions. In this example, I walk you through the construction of this shield step by step and point out various techniques for soldering along the way

Assembly

To assemble this kit, you need to solder the header pins and the pushbutton. Soldering these pieces allows the shield to sit in the header sockets on your Arduino, extending all the pin connections to the proto shield. Note that some versions of the proto board have header sockets (or stackable headers) rather than header pins. Header sockets have long legs so that they can sit on top of an Arduino in the same way as header pins but also allow sockets for another shield to be placed on top. The benefit of header pins is that your shield is shorter and needs less space for any enclosure. Header sockets are used in the assembled shield shown in Figure 9-12. They are the black sockets that run down either side on the top of the board, with pins extending underneath.

In this example, I use header pins and do not connect the ICSP (in-circuit serial programming) connector, which is the 3x2 connector in the center right of the Uno (refer to the first figure in Chapter 2). The ICSP is used as an alternative for uploading sketches with an external programmer as opposed to the Arduino and is for advanced users.

Header pins

First you will need to cut the header pins to length (Figure 9-14). This kit uses a length of 1 x 40, which is 1 row of 40 pins. The plastic strip has a notch between each pair of pins that you can cut to divide the pins neatly. To secure the shield, you need lengths of 6 pins (for the analog-in pins), 8 pins (for the power pins), 8 pins (for the shorter row of digital pins), and 10 pins (for the longer row of digital pins). Use your clippers to cut the header pins to the correct length; you should have 8 pins remaining. (Put these in a box for future use!) The pins should fit exactly because there is a 2.54mm (0.1 inch) pitch between them, which matches the board. You need to look for this same pitch if you're buying header pins of any other connectors separately.

FIGURE 9-14: Close up of the header pins.

Now that you know where the header pins go, you can solder them in place. In the next section, I talk you through soldering technique. Have a read through before you start.

Acquiring Your Soldering Technique

The first step with soldering is to make sure that your components are secure. It's common to try balancing your circuit board on whatever objects are closest to hand to arrange it in an accessible position. But if you do, the board is destined to fall over at some point, most likely when you have a hot soldering iron in your hand.

As mentioned in the "Gathering What You Need for Soldering" section, earlier in this chapter, two good ways to secure your work are to use a third hand or adhesive putty. You can use the crocodile clips on the third hand to grip the plastic of the pins and the circuit board as well, holding them firmly at 90 degrees from one another. Similarly, you can hold the pins at 90 degrees and press adhesive putty into the underside to hold them together. You can then press them into a rigid, weighty support to bring the board closer to you. I sometimes use a big reel of solder for the support.

WARNING

A third way to secure your components might occur to you, and because it's tempting, I describe it here so that you won't do it! You might think that the perfect way to lay out the pins would be to place them in the Arduino itself (with the long ends sticking into the sockets) and then place the circuit board on top. This approach holds everything at 90 degrees; however, because the pins are designed to connect to the Arduino and conduct electrical current, they can also conduct other things, such as heat. If they conduct the heat of your soldering iron, that heat can be passed through the board to the legs of the very sensitive microcontroller chip and damage it irreparably.

When you have your circuit board supported and ready, you can rotate it to a comfortable working angle, most likely the same angle as your iron, as shown in Figure 9-15.

Fire up your soldering iron. My Weller WES51 has a temperature range of 35–85 and the units show this as °F x 10, so the real range is 350–850 °F! The hotter you set it, the quicker it will melt solder to make joints, but it will also melt everything else faster, such as plastic parts and silicon chips. Always set your iron to the lowest convenient temperature. A good way to test the temperature of the iron is to melt some solder. If it's taking a long time to melt, make sure that you have a good amount of the surface area of the soldering iron's tip in contact with the solder. If it's still not working, increase the temperature gradually, allowing time for it to get up to temperature. I normally set it to 650 °F (340 °C), which is hot enough but not too hot.

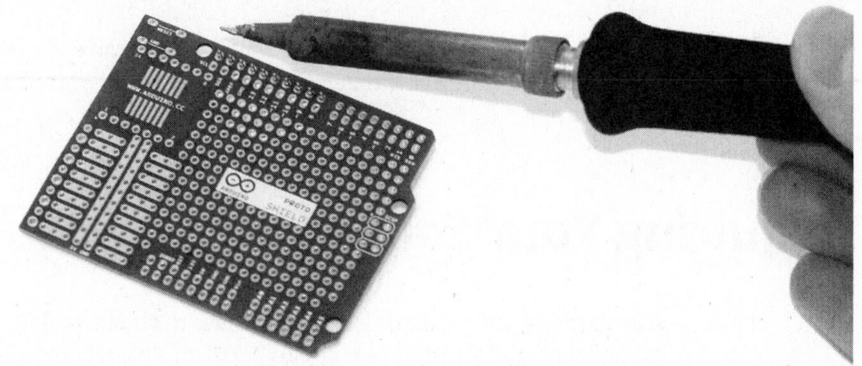

FIGURE 9-15: Always arrange your work in a comfortable position.

Some quality soldering irons have a thermostat that tells you when the iron is up to temperature. The really fancy ones have a digital display. With the cheaper ones, you have to use your best judgment.

While your iron is getting up to temperature, you can wet your sponge. On some irons, the sponge is stuck down for convenience, but it's particularly inconvenient when you need to wet it. I recommend unsticking it and taking it over to a basin rather than spilling water over your soldering iron and the surrounding area. The sponge should be damp but not full of water. The dampness stops the sponge from burning when you pass your iron across it. If the sponge is too wet, however, it can lower the temperature of the iron and the solder, meaning that the solder hardens or solidifies and can't be removed until it's up to temperature again.

Now you're ready you can start soldering. Follow these steps:

1. **Melt a small amount of solder on the tip of your iron (called *tinning the tip*); see Figure 9-16.**

   The solder should stick to the edge and smoke. Generally, with a new or well-maintained iron, the solder latches onto the tip with no problem. If it doesn't stick to the edge of the tip, you may need to try rotating it to find a good patch that isn't oxidized. Failing that, you can use some tip cleaner to remove any built-up layer. By pressing your hot iron into the tip cleaner and wiping away any buildup as it loosens, you can restore your iron to its former glory.

WARNING

Tip cleaner is generally a nasty, toxic substance. Make sure you do not ingest or inhale any of it.

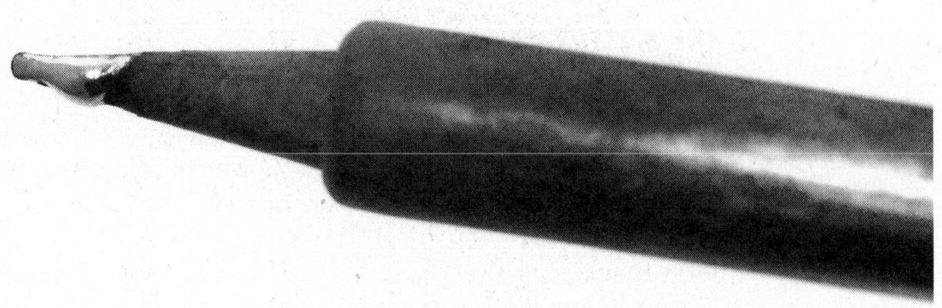

FIGURE 9-16: Tinning the tip helps to preserve your iron and makes soldering easier.

2. **When you have a blob of solder on your iron, wipe it off on a sponge to reveal a bright metallic tip on your iron.**

   The aim is to apply this freshly tinned edge (not the point) to the area that you're joining. Using the edge gives you more surface area and allows the joint to heat up faster. It's also important to note that a variety of tips are available for soldering irons, and if you choose good ones, they are easily interchangeable, allowing you to suit them to different situations. Some are pointed, as in Figure 9-16; others are screwdriver tips, chisel tips, and beveled tips to provide different amounts of surface area for different situations.

3. **Starting on the first pin in one of the rows, apply the iron to the metal plate on the circuit board and the pin connecting to it.**

   This step heats the board and pin, preparing them for the solder.

4. **With your other hand (the one not holding the iron), apply the solder to the point where the iron, the pin, and the metal plate all meet.**

   As the solder is heated, it melts and spreads to fill the gap, sealing the pin and the board together. To apply more solder, press it into the joint where it is melting; to apply less, simply pull it back out. You may need only a few millimeters of solder for small joints.

5. **When the solder has filled the area, remove it, but keep your iron there for a second or two longer.**

   This extra time allows the solder to melt fully and fill any gaps for a solid joint.

6. **Remove your soldering iron by wiping it up the leg of the pin.**

   Any excess solder is directed upward into a point that can be cut off rather than sitting in a blob.

This entire process should take around 2–3 seconds. That probably sounds impossible, but after you get used to the rhythm, it's totally achievable. All it requires is practice.

Following the preceding steps should leave you with a neat, pyramid-shaped solder joint, with the metal pad on the circuit board completely covered. If you can see the hole that the pin has come through or have a blob of solder on the pin that is not connected to the circuit board, you need to reapply heat using the soldering iron, and maybe reapply a little solder as well.

After you solder the first pin, it's a good idea to do the pin at the other end of the row. By doing these two, you secure the row in place, and if it is not level or at 90 degrees, you can still straighten the row by heating up the solder at either end. If you find that you have too much solder, first try reapplying heat. Watch as the solder melts, and you should see it filling all the gaps. If you still have too much solder, you can use a solder sucker or copper braid (described in "Gathering What You Need to Solder," earlier in this chapter) to remove any excess and try again.

For examples of well-soldered joints versus bad ones, see the image at `learn.sparkfun.com/tutorials/how-to-solder-through-hole-soldering#soldering-your-first-component-`.

When you are happy with the row, you can solder every pin in place and repeat this process for each section of header pins. For the pushbutton, simply place it in the position indicated on the top side of the circuit board; it should clip in because of the design of the pushbutton legs. Secure with adhesive putty if necessary and then turn the board over and repeat the same soldering technique as before. When you are finished, do one last visual check to make sure that no solder joints are touching. Having two pins connected could cause a short circuit and damage your board, so check carefully. If in doubt, you can use the continuity test function on your multimeter to check consecutive pins. You shouldn't hear any beeps on pins that should not be joined.

Place the shield on top of your Arduino and check that it fits correctly. If it does, you're ready to solder your first circuit, which you find out about next.

Building Your Circuit

After you have a shield to work on (see "Assembling a Shield," earlier in this chapter), you can think about what circuit you want to build on it. Before you actually build that circuit, though, you should do some proper planning to avoid having to undo what you've done, which can be difficult. The best way to start is by prototyping the circuit on a solder-less breadboard. This process is quick, easy, and — more importantly — not permanent. As covered in Chapters 6 and 7, throwing together a circuit and making sure that it works are simple tasks. For this example, I use the AnalogInOutSerial example (see Chapter 6) to demonstrate how you can transform a solderless breadboard into a soldered one.

Knowing your circuit

First, recreate the AnalogInOutSerial circuit on the breadboard as shown at the end of Chapter 6. Upload the sketch by choosing File⇨Examples⇨03.Analog⇨AnalogInOutSerial. This should give you an LED that can be faded by twisting the potentiometer.

When you have this circuit working, take a look at the AnalogInOutSerial circuit again on the breadboard. One difference that you can immediately see is that the proto shield doesn't have the same rows and columns as the breadboard, except in one corner. This corner area is designed for ICs (integrated circuits) or chips but can be used for anything. The rest of the proto shield has individual holes into which components can be soldered. You can connect these with wires to join the various components.

The easiest way to convert the circuit to one that will work on the shield is to look at the circuit diagram. The lines that connect the components can be substituted for real wires and soldered directly to the correct pins. For the potentiometer, you need three wires: 5V, GND, and analog 0. For the LED and resistor, you need two more: GND and pin 9. It's always a good idea to draw the circuit first for clarity. It doesn't need to be as neat as the circuit diagram in the examples, but planning your circuit will save a lot of painful de-soldering later. As the old carpentry saying goes, *measure twice, cut once* (or in this case, *sketch twice, solder once)*.

Note that the wires all go to holes next to the wire with which they need to connect. Because you don't have enough space to comfortably fit the wires and the component legs in one hole, you need to get them close and then bridge the gap by using the ends of the wire or the legs of the components.

Laying out your circuit

Now that you've drawn the circuit, you should lay it out so that you know what length to cut your wires. To secure the components to the board, insert them and bend to 45 degrees the parts of the legs that protrude through the board. They should look something like that shown in Figure 9-17.

FIGURE 9-17: An LED secured in a circuit board, seen from below.

Preparing your wire

You can see that the lengths of wire required are relatively short. If you have solid-core wire, you can bend it neatly into shape to sit flat on the circuit board by using needle-nose pliers. If you have multicore wire, making an arc of wire up out of one hole and into the next is easier. Remember to measure or estimate the distance and add a small amount on each end to fit into the hole before you cut your lengths. Strip back the wire and sit it in place to make sure that the length is correct. Note that sometimes the end of multicore wire can become frayed. If this happens, grip it between your thumb and forefinger and twist the end into a point. If you lightly coat the end in solder, you can prevent the wire from unravelling and tin it in preparation for soldering. When you're happy with the length, place the wires to one side.

Soldering your circuit

Now that you have all the components and wires, it's time to solder them in place. Earlier, you bent the component leg at 45 degrees, so the wires should still be hanging on. Because the resistor and LED are next to each other, you can use their legs to connect them, avoiding the use of another wire. Remember to check that your LED is the right way around. If required, use adhesive putty to further secure the components in place and then solder them as with the pin headers.

When all the components are in place, you can solder the wire. Insert the wire and bend to 45 degrees as before. When it is secured, solder it in place or bend it further to meet the component legs, bridging the gap as with the resistor and LED. As always, there is no right or wrong in the appearance of your result; it depends how neat you want to be. Some people prefer to wind the wire around the legs of the component to get a good grip; others prefer to join them side by side to get a clean joint. The choice is yours. As long as you have a good connection, you're in good shape.

Cleaning up

When you finish soldering, give the board a good check for any loose connections. If all seems well, you can start neatening the board. Using your clippers, carefully remove the legs of the components from just above the solder joint, at the top of the pyramid. You can cut lower, but the solder is thicker and you risk tearing the metal contacts off the circuit board, which cannot be fixed. Remember, always hold or cover the piece of metal that you're clipping. If you don't, it can fly a great distance and seriously injure someone.

Testing your shield

Now that you have your shield with a completed circuit assembled on it, it's time to plug it in and try it. If everything is working correctly, you should have a neatly packaged dimmer circuit shield resembling something like Figure 9-18.

FIGURE 9-18: My new dimmer shield, ready to be taken out into the world!

Packaging Your Project

Now that your circuit is no longer at risk of falling apart, it's a good idea to look at protecting it from the outside world by boxing it up.

Enclosures

The simplest way to protect your circuit is by putting it in a box. In electronics terms, such a box is called an enclosure or a project box. You can find a variety of plastic or metal enclosures in a vast array of shapes, sizes, and finishes. The only task is to find the one that's right for you.

Many of the online suppliers (RS, Farnell, Digi-Key, and others) have huge lists of possible enclosures, but you can't tell whether it's right without holding it in your hands. Many enclosures have accurate measurements for internal and external dimensions, but even with that information there are usually omissions, such as the molded plastic for the screw fixings. My advice is to find a retail store, such as Radio Shack, and take an Arduino with you to see whether it will fit correctly with enough space for wires and any shields. Then you'll know what works and what to order next time.

Keep the following considerations in mind when boxing up your project.

- **The ability to access the USB for code changes:** You may have to unscrew your enclosure to update the code. If this process is too time consuming, you may need to drill a hole big enough to plug the USB in from the outside.
- **Power to run the Arduino:** If your Arduino is not powered by USB, how is it powered? It could be an external power supply that plugs into the power jack, which needs a big enough hole for the plug. You could remove the plug and solder the bare wires to the Vin and Gnd pins if you're looking for something more permanent. Or you could even run it off a battery pack and only open it every week or so to charge it.
- **The ability to access the inputs and outputs:** What use is an LED or a button if it's inside the box? Most projects need some contact with the outside world, if only to tell you that they're still working, and most components are designed with this need in mind. The lip on an LED means that you can drill a 1.9-inch (5 mm) hole and push the front end through without it going all the way; if you take the plastic or metal knob off a radio, you'll see a similar fitting too.

Always think carefully about the needs of your circuit before soldering it in place and boxing it up. Examine other cheap electronics around you to see what tricks the industry has been using. For example, you may be surprised to find out that most remote controls for remote control cars that give you forward, backward, left, and right are just a few simple pushbuttons underneath those complex-looking control sticks.

Wiring

To give you more flexibility with your wiring, consider attaching inputs, outputs, or power to flexible lengths of cable by using terminal blocks, which are sometimes also known as connector strips, screw terminals, or chocolate (choc) blocks. By doing this, you can fix the inputs and outputs to the enclosure rather than to the circuit board. This approach gives you more flexibility — and if you drill a hole slightly out of alignment, you won't need to de-solder or re-solder your circuit.

Following is a little more detail on adding wires to your project and selecting terminal blocks, which adds flexibility to your project and makes assembling and disassembling your project a lot easier.

Twisting and braiding

When you make wire connectors, I recommend twisting or braiding the wires together. The braiding gives the wires extra strength if pulled on and, as a bonus, looks nice! To twist two wires together, cut them to length and grip one end in a power drill. Hold onto the other end of the two wires with your hand or a vice. Pull the wire taut enough to have no slack but not too tight. Spin the drill until the wires are neatly twisted together, and you should be left with one twisted wire.

If you have three wires, you can braid them in the same way that you braid hair. With three wires in your hand, hold them at one end facing forward. Pass the leftmost wire over the middle and under the rightmost. Keep repeating this process with the new left wire until you have braided the full length. Your project will be more robust and look extra professional. I advise you to use different colors for the wires; if you use the same color wire for all three, you need to use the continuity tester on your multimeter to figure out which is which.

Terminal blocks

Terminal blocks come in a variety of sizes, depending on the amount of current passing through them, and are usually marked with the upper limit that they can maintain. When selecting one for your power supply, always read the current rating and choose a size with a bit of tolerance. If you have a 3A supply, choose a 5A terminal block. When connecting wires, especially multicore wires, you should tin the tip of the wire with a small amount of solder or fold the wire back under the insulation layer so that the screw grips the insulated side. This technique prevents the screw from cutting any of the strands of wire as it is tightened.

Securing the board and other elements

When you're happy with your cabling and have all the required holes, it's a good idea to secure your items so that they don't rattle around inside the box. To secure your Arduino, screw terminals, or stripboard, you can use Velcro-type tape or hot glue. If you have any loose wires, you can use cable ties to neatly tie them together.

» Debouncing your buttons

» Getting more from your buttons

» Averaging your results

» Adjusting the sensitivity of sensors

Chapter 10
Getting Clever with Code

As you find different uses and needs for Arduino, you can refine your code to make it more accurate, responsive, and efficient. Also, by thinking about the code in your project, you may be able to avoid or minimize many of the unexpected results that can occur when dealing with physical hardware and the real world. In this chapter, you look at a few sketches that will help you fine-tune your project.

Blinking Better

Blink is most likely the first sketch you encountered. It's a magical moment when that first LED lights up, isn't it? But what if I told you that it can get even better? The basic Blink sketch presented in Chapter 3 performs its task well, with one significant drawback: It can't do anything else while blinking.

Take a look at the Blink sketch again:

```
void setup() {
  // initialize digital pin LED_BUILTIN as an output.
  pinMode(LED_BUILTIN, OUTPUT);
}
```

```
// the loop function runs over and over again forever
void loop() {
  digitalWrite(LED_BUILTIN, HIGH);   // turn the LED on (HIGH is the voltage
   level)
  delay(1000);                       // wait for a second
  digitalWrite(LED_BUILTIN, LOW);    // turn the LED off by making the
   voltage LOW
  delay(1000);                       // wait for a second
}
```

The loops can be summarized this way:

1. Turn on the LED.
2. Wait for a second.
3. Turn off the LED.
4. Wait for a second.

This delay, or waiting, can be problematic for many people when they try to integrate the Blink sketch with another bit of code. When the sketch uses the delay function, it waits for the amount of time specified (in this case, a second), during which it doesn't do anything else. Effectively, the sketch is twiddling its thumbs.

If you wanted to change something — for example, you wanted the LED to blink only when a light sensor was dark — you might think of writing the code in the loop section something like this:

```
void loop() {

sensorValue = analogRead(sensorPin);

if (sensorValue < darkValue) {
  digitalWrite(LED_BUILTIN, HIGH);   // turn the LED on (HIGH is the voltage
   level)
  delay(1000);                       // wait for a second
  digitalWrite(LED_BUILTIN, LOW);    // turn the LED off by making the
   voltage LOW
  delay(1000);                       // wait for a second
          }
}
```

This code almost works. When the value threshold for `darkValue` is crossed, the `if` loop starts and turns on the LED, waits for one second, and then turns it off and waits for one second. But because the sketch is occupied doing this blink for two seconds, the sketch can't check to see if the light level becomes brighter again until the blink has finished.

The solution is to use a timer rather than pause the program. A timer or counter is like a clock that can be used to time events. For example, the timer can count from 0 to 1,000, and when it reaches 1,000 it can *do something*, and then start counting from 0 again. Timers can be especially useful for regularly timed events, such as checking a sensor every second — or in this case triggering an LED every second.

Setting up the BlinkWithoutDelay sketch

To complete this project, you need:

- » An Arduino Uno
- » An LED

Place the legs of the LED between pin 13 (long leg) and GND (short leg), as shown in Figures 10-1 and 10-2. This placement makes it a bit easier to see the blink in action. If you don't have an LED, look for the one fixed to your Arduino marked L. Upload the sketch to the correct serial port to see the LED blinking away as it would with the standard Blink sketch.

Find the BlinkWithoutDelay sketch by choosing File⇨Examples⇨02.Digital⇨ BlinkWithoutDelay and then open it.

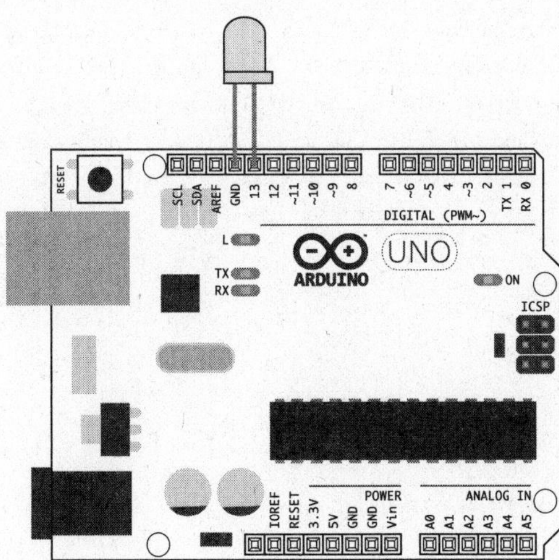

FIGURE 10-1: All you need is an LED in pin 13.

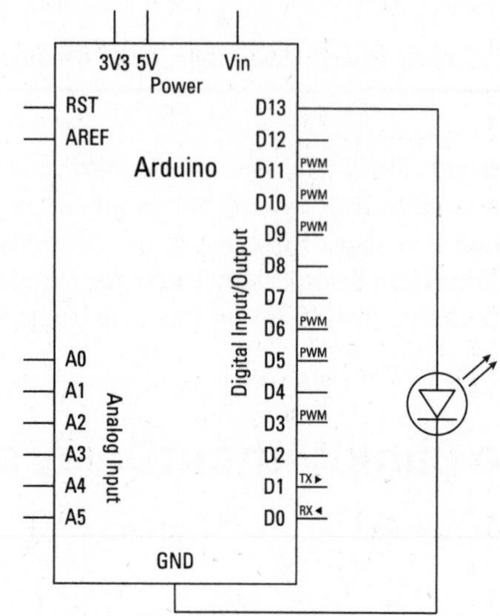

FIGURE 10-2: A circuit diagram showing the LED in pin 13.

The complete code for the BlinkWithoutDelay sketch is as follows:

```
/*
  Blink without Delay

  Turns on and off a light emitting diode (LED) connected to a digital pin,
  without using the delay() function. This means that other code can run at the
  same time without being interrupted by the LED code.

  The circuit:
  - Use the onboard LED.
  - Note: Most Arduinos have an on-board LED you can control. On the UNO, MEGA
    and ZERO it is attached to digital pin 13, on MKR1000 on pin 6. LED_BUILTIN
    is set to the correct LED pin independent of which board is used.
    If you want to know what pin the on-board LED is connected to on your
    Arduino model, check the Technical Specs of your board at:
    https://www.arduino.cc/en/Main/Products

  created 2005
  by David A. Mellis
  modified 8 Feb 2010
  by Paul Stoffregen
  modified 11 Nov 2013
  by Scott Fitzgerald
  modified 9 Jan 2017
  by Arturo Guadalupi
```

```
  This example code is in the public domain.

  http://www.arduino.cc/en/Tutorial/BlinkWithoutDelay
*/

// constants won't change. Used here to set a pin number:
const int ledPin =  LED_BUILTIN;// the number of the LED pin

// Variables will change:
int ledState = LOW;             // ledState used to set the LED

// Generally, you should use "unsigned long" for variables that hold time
// The value will quickly become too large for an int to store
unsigned long previousMillis = 0;        // will store last time LED was updated

// constants won't change:
const long interval = 1000;           // interval at which to blink
   (milliseconds)

void setup() {
  // set the digital pin as output:
  pinMode(ledPin, OUTPUT);
}

void loop() {
  // here is where you'd put code that needs to be running all the time.

  // check to see if it's time to blink the LED; that is, if the difference
  // between the current time and last time you blinked the LED is bigger than
  // the interval at which you want to blink the LED.
  unsigned long currentMillis = millis();

  if (currentMillis - previousMillis >= interval) {
    // save the last time you blinked the LED
    previousMillis = currentMillis;

    // if the LED is off turn it on and vice-versa:
    if (ledState == LOW) {
      ledState = HIGH;
    } else {
      ledState = LOW;
    }
```

```
    // set the LED with the ledState of the variable:
    digitalWrite(ledPin, ledState);
  }
}
```

This sketch is quite a bit longer than Blink and may seem more confusing, so we'll walk through it one line at a time to see what's happening.

Understanding the BlinkWithoutDelay sketch

First, in the declarations, a `const int` is used to set `ledPin` to the built-in LED (also known as pin 13) because it is a constant integer and does not change:

```
const int ledPin =  LED_BUILTIN;
```

Next are the variables. `ledState` is set to `LOW` so that our LED starts the sketch in an off state:

```
int ledState = LOW;
```

Then there is a new variable referred to as a `long` rather than an `int`. See the "Long and Unsigned Long" sidebar later in this chapter for more about `long`s. The first instance, `previousMillis`, stores the time in milliseconds so that you can monitor how much time has passed each time you do a loop:

```
unsigned long previousMillis = 0;
```

The second value, a constant named `interval`, is the time in milliseconds between each blink, which is set to 1000 milliseconds, or 1 second:

```
const long interval = 1000;
```

In the `setup` function, you have only one pin to define as `OUTPUT`. Pin 13 is referred to as `ledPin`, just as it is in the declarations:

```
void setup() {
  // set the digital pin as output:
  pinMode(ledPin, OUTPUT);
}
```

LONG AND UNSIGNED LONG

Longs are for extra-long number storage and can store a value from -2,147,483,648 to 2,147,483,647, whereas an `int` can store only -32,768 to 32,767. When measuring time in milliseconds, you need access to big numbers because every second is stored as 1,000 milliseconds. To get an idea of just how big a `long` value is, imagine the maximum amount of time that it could store: 2,147,483.6 seconds or 35791.4 minutes or 596.5 hours or approximately 24.9 days!

In some cases, you have no need for a negative range, so to avoid unnecessary calculations, use an `unsigned long` instead. An `unsigned long` is similar to a regular `long` but has no negative values, which gives your `unsigned long` a whopping range of 0 to 4,294,967,295.

In the loop, things start to get more complicated. Because the code for the timer can be run at the end of every loop, you can add your own code at the start of the loop so that it doesn't interfere with the timer. Following your code, the timer code begins, which declares another variable: an `unsigned long` to store the current value of the timer in milliseconds. The timer code uses the function `millis()`, which returns the number of milliseconds since the current Arduino program began running. After approximately 50 days, this value resets to 0, but this is more than enough time for most applications.

```
unsigned long currentMillis = millis();
```

TIP

Variables declared inside a loop or other functions are known as *local* variables. They exist only within the function in which they are declared (and other sub-functions contained inside), and cease to exist after the function is completed. They are re-declared the next time the function is called. If you have a variable that needs to be either read or written to by other functions or pieces of code, you should use a *global* variable and declare it at the start of the sketch before the `setup` function.

Next you need to check the current `millis()` value to see how much time has passed. You do so by using a simple `if` loop that subtracts the previous value from the current value to get the difference. If that difference is greater than the interval value, the sketch knows that it's time to blink. It's important that you also tell the code to reset `previousMillis`; otherwise, it will measure the interval only once. This is what setting `previousMillis = currentMillis` does:

```
if(currentMillis - previousMillis > interval) {
  // save the last time you blinked the LED
  previousMillis = currentMillis;
```

Because the LED could already be on or off, the code needs to check the state of the LED to know what to do. The state is stored in `ledState`, so another simple `if` statement can check the state and do the opposite. If `LOW`, make `HIGH`; or if `HIGH`, make `LOW`. The following code updates the variable `ledState`:

```
// if the LED is off turn it on and vice-versa:
    if (ledState == LOW){
      ledState = HIGH;
    } else {
      ledState = LOW;
    }
```

Now, all that is left to do is to write the newly updated state to the LED by using `digitalWrite`:

```
    // set the LED with the ledState of the variable:
    digitalWrite(ledPin, ledState);
}
```

This code allows you to happily blink your LED while performing any number of other functions.

Taking the Bounce Out of Your Button

A strange occurrence that happens with pushbuttons is bouncing. The microcontroller on your Arduino can read a switch thousands of times per second, much faster than you can operate it. This speed ensures that the reading is instantaneous (as far as human perception can tell), but sometimes there is a moment of fuzziness when the contact on a switch is neither fully down nor fully up, which causes it to read on and off rapidly in quick succession until it reaches the correct state. This behavior is called bouncing. To remove this peculiarity, you have to ignore any sudden changes when the switch state changes by using a timer. It's relatively simple and can greatly improve the reliability of your buttons. Using timers works for your inputs as well as outputs, as noted in the preceding section, "Blinking Better."

Setting up the Debounce sketch

Complete the circuit in Figure 10-3 to try out the Debounce sketch. You need:

- An Arduino Uno
- A breadboard

- A pushbutton
- An LED
- A 10k ohm resistor
- Jump wires

Complete the circuit shown in Figures 10-3 and 10-4, using a breadboard to mount the pushbutton part of the circuit. The LED can be inserted straight into pin 13 and its neighboring GND pin.

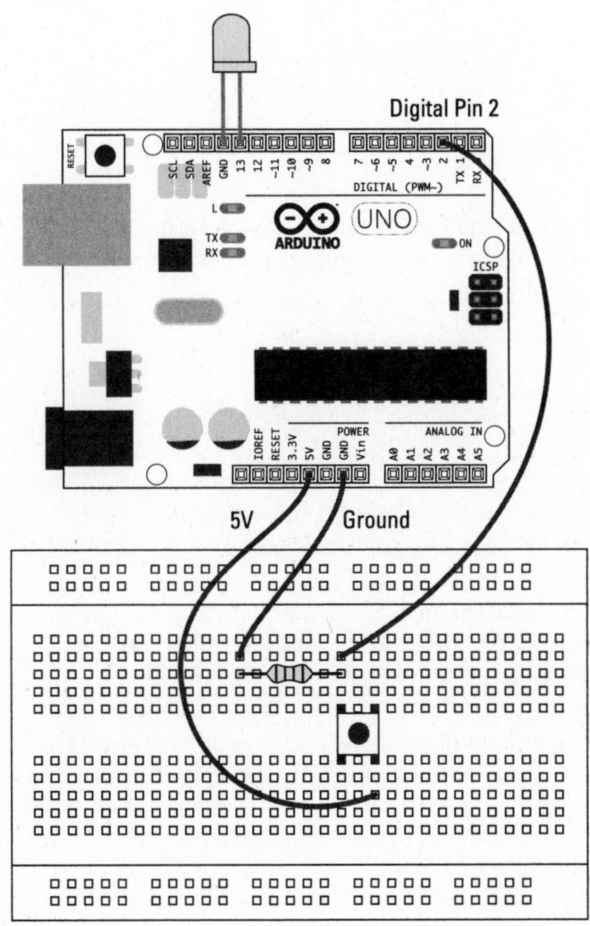

FIGURE 10-3: The pushbutton circuit layout.

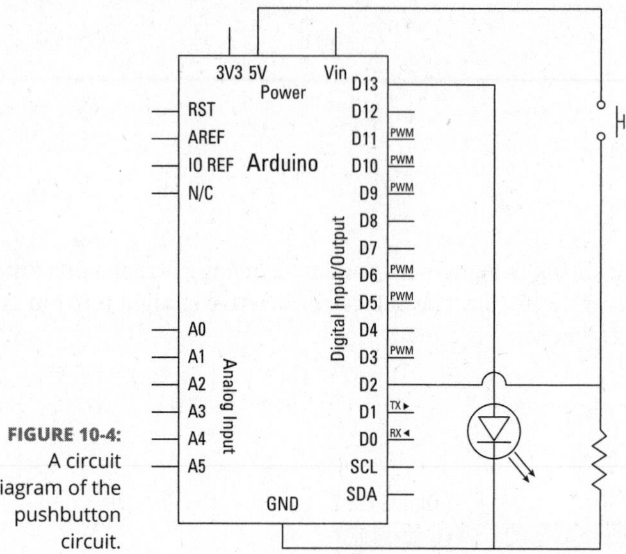

FIGURE 10-4: A circuit diagram of the pushbutton circuit.

Build the circuit and choose File ➪ Examples ➪ 02.Digital ➪ Debounce to find the pushbutton Debounce sketch and open it. The complete code for the Debounce sketch is as follows:

```
/*
  Debounce

  Each time the input pin goes from LOW to HIGH (e.g. because of a push-button
  press), the output pin is toggled from LOW to HIGH or HIGH to LOW. There's a
  minimum delay between toggles to debounce the circuit (i.e. to ignore noise).

  The circuit:
  - LED attached from pin 13 to ground
  - pushbutton attached from pin 2 to +5V
  - 10 kilohm resistor attached from pin 2 to ground

  - Note: On most Arduino boards, there is already an LED on the board connected
    to pin 13, so you don't need any extra components for this example.

  created 21 Nov 2006
  by David A. Mellis
  modified 30 Aug 2011
  by Limor Fried
  modified 28 Dec 2012
  by Mike Walters
  modified 30 Aug 2016
  by Arturo Guadalupi
```

```
  This example code is in the public domain.

  http://www.arduino.cc/en/Tutorial/Debounce
*/

// constants won't change. They're used here to
// set pin numbers:
const int buttonPin = 2;     // the number of the pushbutton pin
const int ledPin =  13;      // the number of the LED pin

// Variables will change:
int ledState = HIGH;         // the current state of the output pin
int buttonState;             // the current reading from the input pin
int lastButtonState = LOW;   // the previous reading from the input pin

// the following variables are long's because the time, measured in miliseconds,
// will quickly become a bigger number than can be stored in an int.
long lastDebounceTime = 0;  // the last time the output pin was toggled
long debounceDelay = 50;    // the debounce time; increase if the output
                            // flickers

void setup() {
  pinMode(buttonPin, INPUT);
  pinMode(ledPin, OUTPUT);
}

void loop() {
  // read the state of the switch into a local variable:
  int reading = digitalRead(buttonPin);

  // check to see if you just pressed the button
  // (i.e. the input went from LOW to HIGH), and you've waited
  // long enough since the last press to ignore any noise:

  // If the switch changed, due to noise or pressing:
  if (reading != lastButtonState) {
    // reset the debouncing timer
    lastDebounceTime = millis();
  }

  if ((millis() - lastDebounceTime) > debounceDelay) {
    // whatever the reading is at, it's been there for longer
    // than the debounce delay, so take it as the actual current state:
    buttonState = reading;
  }
```

```
  // set the LED using the state of the button:
  digitalWrite(ledPin, buttonState);

  // save the reading.  Next time through the loop,
  // it'll be the lastButtonState:
  lastButtonState = reading;
}
```

When you've uploaded the sketch, you should have a reliable, debounced button. It can be difficult to see the effects — when everything's working correctly, you see just accurate button presses and responses from your LED.

Understanding the Debounce sketch

The Debounce sketch starts by declaring two constants that are used to define the input and output pins:

```
// constants won't change. They're used here to
// set pin numbers:
const int buttonPin = 2;     // the number of the pushbutton pin
const int ledPin =  13;      // the number of the LED pin
```

The next set of variables hold details about the button state. `ledState` is set to `HIGH` so that the LED starts as being turned on; `buttonState` is left empty and holds the current state; `lastButtonState` holds the previous button state so that it can be compared with the current state:

```
// Variables will change:
int ledState = HIGH;         // the current state of the output pin
int buttonState;             // the current reading from the input pin
int lastButtonState = LOW;   // the previous reading from the input pin
```

Finally, two `long` variables store time values. These are used in a timer to monitor the time between readings and prevent any sudden changes in values, such as those that occur during bounces:

```
// the following variables are long's because the time, measured in miliseconds,
// will quickly become a bigger number than can be stored in an int.
long lastDebounceTime = 0;  // the last time the output pin was toggled
long debounceDelay = 50;    // the debounce time; increase if the output
                            // flickers
```

The setup function is straightforward and sets only the input and output pins:

```
void setup() {
  pinMode(buttonPin, INPUT);
  pinMode(ledPin, OUTPUT);
}
```

In the loop, a reading is taken from the button pin and stored in the reading variable:

```
void loop() {
  // read the state of the switch into a local variable:
  int reading = digitalRead(buttonPin);
```

The reading variable is then checked against lastButtonState. The first time this runs, lastButtonState is LOW because it was set in the variable declarations at the beginning of the sketch. In the if statement, the comparison symbol != is used. This means: "If reading *is not* equal to lastButtonState, do something." If this change has occurred, lastDebounceTime is updated so that a fresh comparison can be made the next time the loop runs:

```
  // If the switch changed, due to noise or pressing:
  if (reading != lastButtonState) {
    // reset the debouncing timer
    lastDebounceTime = millis();
  }
```

If reading has been the same for longer than the debounce delay of 50 milliseconds, it can be assumed that the value is not erratic and can be forwarded to the buttonState variable:

```
  if ((millis() - lastDebounceTime) > debounceDelay) {
    // whatever the reading is at, it's been there for longer
    // than the debounce delay, so take it as the actual current state:
    buttonState = reading;
  }
```

The trusted value can then be used to trigger the LED directly. In this case, if the button is HIGH, it is closed, so the same HIGH value can be written to the LED to turn it on:

```
  digitalWrite(ledPin, buttonState);
```

The current `buttonState` becomes `lastButtonState` for the next loop, and then the code returns to the start of the `loop` function:

```
  lastButtonState = reading;
}
```

Some pushbuttons and triggers can be more or less reliable than others, depending on the way they're made or used. By using bits of code like this, you can sort out any inconsistencies and create more reliable results.

Making a Better Button

Buttons are usually simple things. They are either on or off, depending on whether or not you're pressing them. You can monitor these changes and interpret them to make a button more intelligent. If you can tell when a button has been pressed, you don't need to constantly read its value and can instead just look for this change of state. This practice is much better when connecting your Arduino to a computer, because it efficiently sends the appropriate data as it is needed, rather than hogging the serial port.

Setting up the StateChangeDetection sketch

To build the StateChangeDetection circuit, you need the following:

- An Arduino Uno
- A breadboard
- A pushbutton
- A 10k ohm resistor
- An LED
- Jump wires

Using the layout and circuit diagrams shown in Figures 10-5 and 10-6, you can lay out a simple button circuit with an LED as an output. The hardware in this circuit is the same as the basic Button sketch, but with the use of some simple code, you can make a button a lot more intelligent.

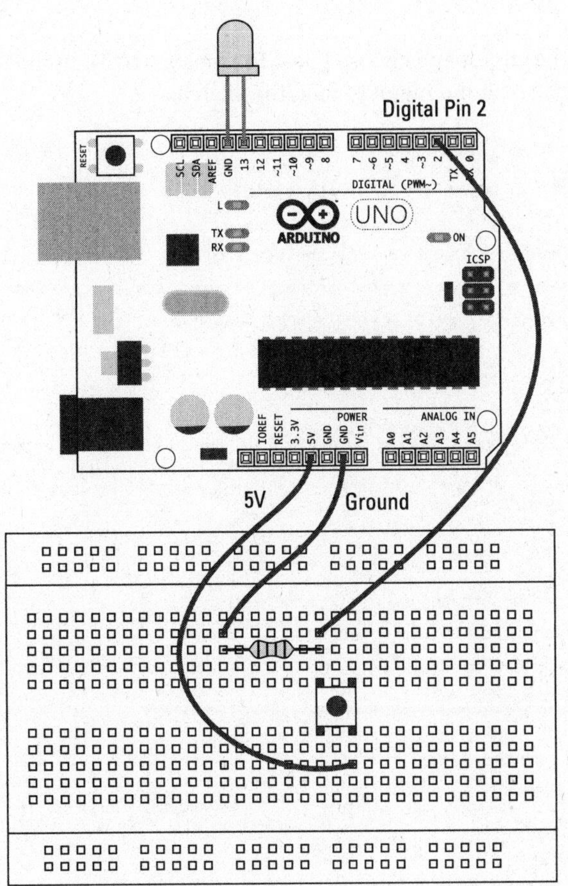

FIGURE 10-5: A button circuit layout.

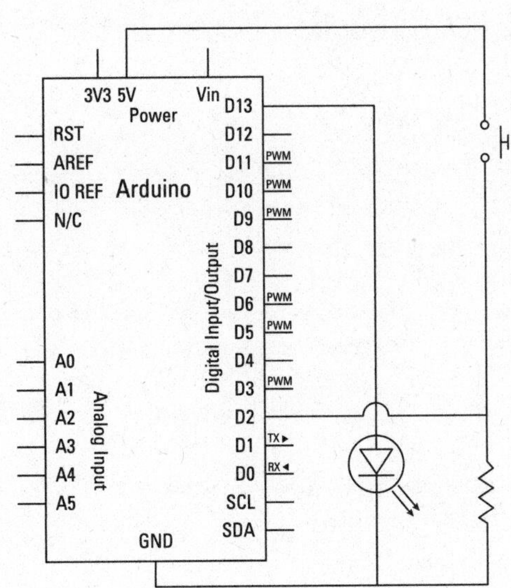

FIGURE 10-6: A button circuit diagram.

Complete the circuit and choose File ➪ Examples ➪ 02.Digital ➪ StateChangeDetection from the Arduino menu to load the sketch:

```
/*
  State change detection (edge detection)

  Often, you don't need to know the state of a digital input all the time, but
  you just need to know when the input changes from one state to another.
  For example, you want to know when a button goes from OFF to ON. This is
   called
  state change detection, or edge detection.

  This example shows how to detect when a button or button changes from off
   to on
  and on to off.

  The circuit:
  - pushbutton attached to pin 2 from +5V
  - 10 kilohm resistor attached to pin 2 from ground
  - LED attached from pin 13 to ground (or use the built-in LED on most
    Arduino boards)

  created  27 Sep 2005
  modified 30 Aug 2011
  by Tom Igoe

  This example code is in the public domain.

  http://www.arduino.cc/en/Tutorial/ButtonStateChange
*/

// this constant won't change:
const int buttonPin = 2;  // the pin that the pushbutton is attached to
const int ledPin = 13;    // the pin that the LED is attached to

// Variables will change:
int buttonPushCounter = 0;  // counter for the number of button presses
int buttonState = 0;        // current state of the button
int lastButtonState = 0;    // previous state of the button

void setup() {
 // initialize the button pin as a input:
 pinMode(buttonPin, INPUT);
```

```
  // initialize the LED as an output:
  pinMode(ledPin, OUTPUT);
  // initialize serial communication:
  Serial.begin(9600);
}


void loop() {
  // read the pushbutton input pin:
  buttonState = digitalRead(buttonPin);

  // compare the buttonState to its previous state
  if (buttonState != lastButtonState) {
    // if the state has changed, increment the counter
    if (buttonState == HIGH) {
      // if the current state is HIGH then the button
      // wend from off to on:
      buttonPushCounter++;
      Serial.println("on");
      Serial.print("number of button pushes: ");
      Serial.println(buttonPushCounter);
    }
    else {
      // if the current state is LOW then the button went from on to off:
      Serial.println("off");
    }
  }
  // save the current state as the last state, for next time through the loop
  lastButtonState = buttonState;


  // turns on the LED every four button pushes by
  // checking the modulo of the button push counter.
  // the modulo function gives you the remainder of
  // the division of two numbers:
  if (buttonPushCounter % 4 == 0) {
    digitalWrite(ledPin, HIGH);
  } else {
  digitalWrite(ledPin, LOW);
  }

}
```

Press the Compile button to check your code. Compiling should highlight any grammatical errors and display a red text box if any are discovered. If the sketch compiles correctly, click Upload to send the sketch to your board. When the sketch has finished uploading, choose the serial monitor, and you should be presented with a readout showing when the button was turned on and off as well as how many times it was pressed. Also, the LED should illuminate every four button pushes to show that it's counting.

If nothing happens, try the following:

- Double-check your wiring.
- Make sure that you're using the correct pin number.
- Check the connections on the breadboard. If the jump wires or components are not connected using the correct rows in the breadboard, they will not work.

Understanding the StateChangeDetection sketch

In the StateChangeDetection sketch, the first action is to declare the constants for the sketch. The input and output pins won't change, so they are declared as constant integers, pin 2 for the pushbutton and pin 13 for the LED.\:

```
// this constant won't change:
const int buttonPin = 2;   // the pin that the pushbutton is attached to
const int ledPin = 13;     // the pin that the LED is attached to
```

Other variables are needed to keep track of the pushbutton's behavior. One variable is a counter that keeps a running total of the number of button presses, and two other variables track the current and previous states of the pushbutton. These are used to monitor the button presses as the signal goes from `HIGH` to `LOW` or `LOW` to `HIGH`:

```
// Variables will change:
int buttonPushCounter = 0;   // counter for the number of button presses
int buttonState = 0;         // current state of the button
int lastButtonState = 0;     // previous state of the button
```

In the `setup` function, the pins are set to `INPUT` and `OUTPUT` accordingly. The serial port is opened for communication to display changes in the pushbutton:

```
void setup() {
  // initialize the button pin as a input:
  pinMode(buttonPin, INPUT);
```

```
  // initialize the LED as an output:
  pinMode(ledPin, OUTPUT);
  // initialize serial communication:
  Serial.begin(9600);
}
```

The first stage in the main loop is to read the state of the pushbutton:

```
void loop() {
  // read the pushbutton input pin:
  buttonState = digitalRead(buttonPin);
```

If this state is not equal to the previous value, which happens when the pushbutton is pressed, the program progresses to the next `if ()` statement.

```
  // compare the buttonState to its previous state
  if (buttonState != lastButtonState) {
```

The next condition is to check whether the button is `HIGH` or `LOW`. If it is `HIGH`, the pushbutton has changed to on:

```
    // if the state has changed, increment the counter
    if (buttonState == HIGH) {
      // if the current state is HIGH then the button
      // wend from off to on:
```

This bit of code increments the counter and then prints a line on the serial monitor to show the state and the number of pushes. The counter is incremented on the downward part of the button press rather than the release:

```
      buttonPushCounter++;
      Serial.println("on");
      Serial.print("number of button pushes: ");
      Serial.println(buttonPushCounter);
    }
```

If the pushbutton went from `HIGH` to `LOW`, the button state is off, and this change of state is printed on the serial monitor, as shown in Figure 10-7.

```
    else {
      // if the current state is LOW then the button
      // went from on to off:
      Serial.println("off");
    }
  }
```

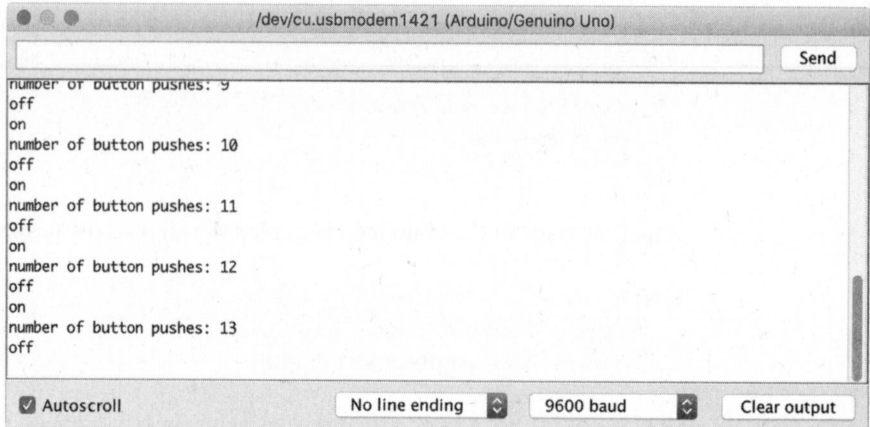

FIGURE 10-7: The serial monitor provides a window into what your Arduino is experiencing.

This piece of code allows you to click a pushbutton rather than hold down on it; the code also leaves plenty of room to add your own functionality.

Because a change has occurred, the current state becomes the last state in preparation for the next loop:

```
// save the current state as the last state,
//for next time through the loop
lastButtonState = buttonState;
```

At the end of the loop, a check is done to make sure that four button presses have occurred. If the total number of presses divided by four gives a remainder of 0, the LED pin is set to `HIGH`; if not the pin is set to `LOW` again:

```
// turns on the LED every four button pushes by
// checking the modulo of the button push counter.
// the modulo function gives you the remainder of
// the division of two numbers:
if (buttonPushCounter % 4 == 0) {
 digitalWrite(ledPin, HIGH);
} else {
digitalWrite(ledPin, LOW);
}

}
```

TIP

You may find that the counter in this sketch can occasionally jump up, depending on the type and quality of pushbutton that you're using, causing you to quite rightly ask, "Why doesn't this sketch include debouncing?" The sketches in the Arduino examples are designed to help you understand many individual principles easily, to equip you for any situation. To include two or more techniques in one sketch may be good for the application, but would make it more difficult for you, the person learning, to understand how each element works.

It is possible to combine numerous examples to reap the benefits of each, but this topic is beyond the scope of this book. If you want to try your hand at including two techniques in a single sketch, open two sketch examples side by side and combine them into one sketch, one line at a time, checking as you go that there are no repetitions of variables or omissions. The Compile shortcut (Ctrl+R or cmd+R) is helpful for this task. Good luck!

Smoothing Your Sensors

Analog sensors measure light levels or distance to a high degree of accuracy. Sometimes, however, they can be overly sensitive and flinch at the slightest change. If that's what you're looking for, great. If not, you may want to smooth the results to compensate for any erroneous readings. Smoothing is effectively averaging your results so that any of these anomalies don't affect your reading as much. In Chapter 14 you also learn about showing these results on a bar graph, using Processing, which can be a great help for spotting inconsistencies.

Setting up the Smoothing sketch

For this example, you try smoothing on a light sensor.

For the Smoothing sketch, you need the following:

- An Arduino Uno
- A breadboard
- An LED
- A light sensor
- A 10k ohm resistor
- A 220 ohm resistor
- Jump wires

Complete the circuit for reading a light-dependent resistor (LDR) as shown in Figures 10-8 and 10-9.

Choose File ⇨ Examples ⇨ 03.Analog ⇨ Smoothing to find the sketch and upload it.

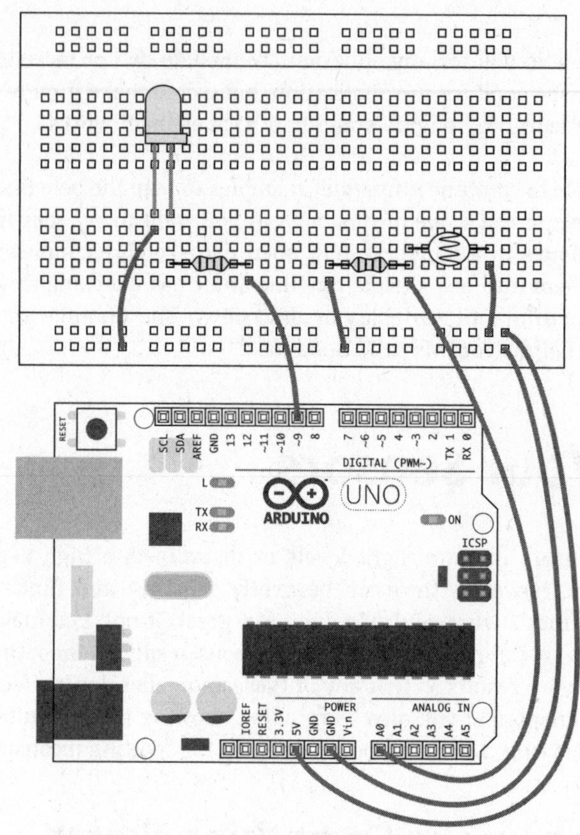

FIGURE 10-8: The light sensor circuit layout.

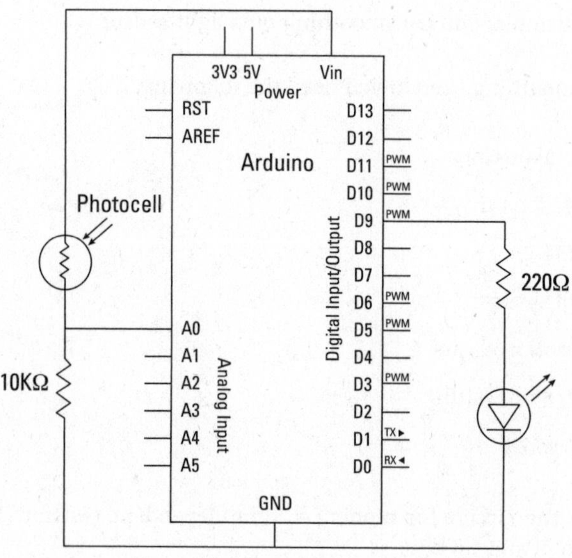

FIGURE 10-9: A circuit diagram of the light sensor circuit.

TIP

The sketch indicates that a potentiometer works for testing. This is true, but it is more difficult to see the effects of smoothing with a potentiometer because the mechanics of the device already provide a smooth analog input. Light, distance, and movement sensors are far more likely to need smoothing.

```
/*
Smoothing

  Reads repeatedly from an analog input, calculating a running average and
  printing it to the computer. Keeps ten readings in an array and continually
  averages them.

  The circuit:
  - analog sensor (potentiometer will do) attached to analog input 0

  created 22 Apr 2007
  by David A. Mellis  <dam@mellis.org>
  modified 9 Apr 2012
  by Tom Igoe

  This example code is in the public domain.

  http://www.arduino.cc/en/Tutorial/Smoothing
*/

// Define the number of samples to keep track of.  The higher the number,
// the more the readings will be smoothed, but the slower the output will
// respond to the input.  Using a constant rather than a normal variable lets
// use this value to determine the size of the readings array.
const int numReadings = 10;
int readings[numReadings];      // the readings from the analog input
int readIndex = 0;              // the index of the current reading
int total = 0;                  // the running total
int average = 0;                // the average

int inputPin = A0;

void setup() {
  // initialize serial communication with computer:
  Serial.begin(9600);
  // initialize all the readings to 0:
  for (int thisReading = 0; thisReading < numReadings; thisReading++)
    readings[thisReading] = 0;
}
```

```
void loop() {
  // subtract the last reading:
  total= total - readings[index];
  // read from the sensor:
  readings[index] = analogRead(inputPin);
  // add the reading to the total:
  total= total + readings[index];
  // advance to the next position in the array:
  index = index + 1;

  // if we're at the end of the array...
  if (index >= numReadings)
    // ...wrap around to the beginning:
    index = 0;

  // calculate the average:
  average = total / numReadings;
  // send it to the computer as ASCII digits
  Serial.println(average);
  delay(1);        // delay in between reads for stability
}
```

This sketch gives you a nicely smoothed reading based on what the sensor is detecting. The smoothing is achieved by averaging a number of readings. The averaging process may slow the number of readings per second, but because the Arduino is capable of reading these changes far faster than you are, the slowness doesn't affect how the sensor works in any noticeable way.

Understanding the Smoothing sketch

The start of the Smoothing sketch declares the constants and variables. First is the number of readings to average, declared as `numReadings` with a value of 10:

```
const int numReadings = 10;
```

The next four variables keep track of how many readings have been stored and average them. These sensor readings are added to an array (or list), which is defined here as `readings`. The number of items in the `readings` array is defined in the square brackets. Because `numReadings` has already been declared, it can be used to set the array length to 10 (which is numbered, or indexed, from 0 to 9):

```
int readings[numReadings];      // the readings from the analog input
```

Index is the common term for the current value and is used to keep track of how many loops or readings are taken. Because the index increases every time a reading is taken, it can be used to store the results of that reading in the correct place in your array, before increasing to store the next reading in the next position in the array:

```
int readIndex = 0;                  // the index of the current reading
```

The `total` variable provides a running total that is added to as readings are made. When the total is processed, the average value is stored in the `average` variable:

```
int total = 0;                  // the running total
int average = 0;                // the average
```

The last variable is `inputPin`, the analog in pin that is being read:

```
int inputPin = A0;
```

In the `setup` function, the serial port is initialized to allow you to view the readings from the light sensor:

```
void setup()
{
  // initialize serial communication with computer:
  Serial.begin(9600);
```

Next in the code is a `for` loop, which is used to effectively reset the array. In the loop, `thisReading`, a new local variable, is initialized and made equal to zero. The `thisReading` variable is then compared to the length of the array. If it's less than the length of the array, the current reading of the value in that part of the array is made equal to zero:

```
// initialize all the readings to 0:
  for (int thisReading = 0; thisReading < numReadings; thisReading++)
    readings[thisReading] = 0;
}
```

In layman's terms, the code reads like this: "Make a variable equal to 0, and if that variable is less than 10, make that same value in the array equal to 0; then increase the variable by 1." As you can see, the code works through all the numbers (0 to 9) and sets that same position in the array to a 0 value. After it reaches 10, the `for` loop ceases running and the code moves on to the main loop.

TIP

This type of automation is great for setting arrays. The alternative is to write all the numbers in the array as individual integer variables, which is a lot less efficient, both for you and the Arduino.

The first line of code in the main loop subtracts any reading in the current index of the array from the total. That value is replaced in this loop, so it is essential to remove it from the total first:

```
void loop() {
  // subtract the last reading:
  total= total - readings[index];
```

The next line obtains a new reading using `analogRead`, which is stored in the current index of the array, overwriting the previous value:

```
  // read from the sensor:
  readings[index] = analogRead(inputPin);
```

This reading is then added to the total to correct it:

```
  // add the reading to the total:
  total= total + readings[index];
```

```
  // advance to the next position in the array:
  index = index + 1;
```

It's important to check when the end of the array is reached so that the program doesn't loop forever without telling you your results. You can do this check with a simple `if` statement: If the index value is greater than or equal to the number of readings that the sketch is looking for, set `index` back to 0. This `if` statement counts the index value from 0 to 9, as in `setup`, and then resets as soon as it reaches 10:

```
// if we're at the end of the array...
  if (index >= numReadings)
    // ...wrap around to the beginning:
    index = 0;
```

To get the average from all the data in the array, the total is simply divided by the number of readings. The average is then displayed on the serial monitor for you to check. Because of the command used to display the message, this action could also be referred to as "printing to the serial port." A 1-millisecond delay at the end slows the program considerably as well as helps to prevent erratic readings:

```
  // calculate the average:
  average = total / numReadings;
```

```
  // send it to the computer as ASCII digits
  Serial.println(average);
  delay(1);        // delay in between reads for stability
}
```

Using simple procedures like this to average your results helps control unpredictable behavior in your projects. Averaging is especially useful if the sensor readings are directly linked to your output.

Calibrating Your Inputs

Think of calibrating your circuit as setting the thermostat in your home. Your furnace or boiler is capable of a range of temperatures, but depending on where you are in the world, different temperatures are appropriate. If you're in a mild climate, you may have the heating on only infrequently for a few months, but if you're in a cold climate, you may have the heating on every night for most of the year.

By calibrating the sensors on your Arduino project, you can tailor the sensor to its location. In this example, you learn how to calibrate a light sensor. Light, of course, is highly variable, whether you're inside, outside, in a well-lit room, or working by candlelight. Despite the huge variation, all these ranges of light can be sensed and interpreted by your Arduino as long as it knows the range. The following sketch shows you how to calibrate a light sensor to its surroundings.

Setting up the Calibration sketch

To try the Calibration example, complete the circuit shown in Figures 10-10 and 10-11 to calibrate your light sensor automatically.

You need:

- An Arduino Uno
- A breadboard
- An LED
- A light sensor
- A 10k ohm resistor
- A 220 ohm resistor
- Jump wires

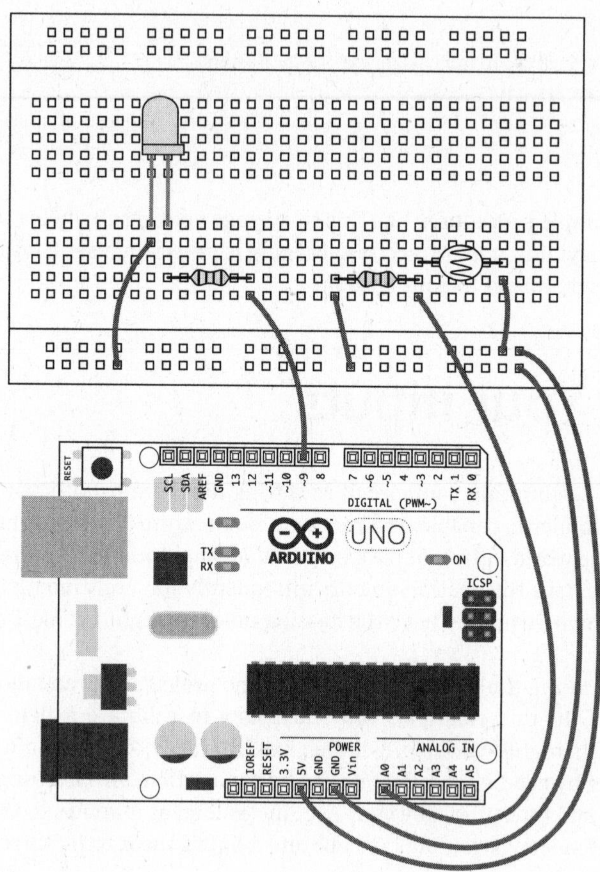

FIGURE 10-10: The light sensor circuit layout.

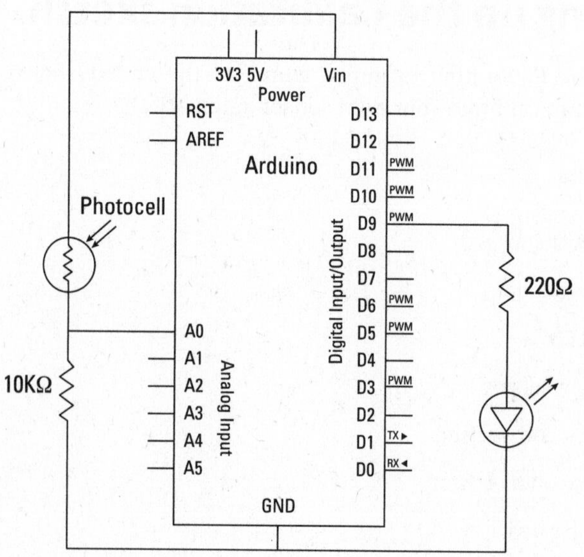

FIGURE 10-11: A circuit diagram of the light sensor circuit.

Build the circuit and go to File⇨Examples⇨03.Analog⇨Calibration to find the sketch. The code for this example is as follows:

```
/*
Calibration

Demonstrates one technique for calibrating sensor input. The sensor readings
during the first five seconds of the sketch execution define the minimum and
maximum of expected values attached to the sensor pin.

The sensor minimum and maximum initial values may seem backwards. Initially,
you set the minimum high and listen for anything lower, saving it as the new
minimum. Likewise, you set the maximum low and listen for anything higher as
the new maximum.

The circuit:
- analog sensor (potentiometer will do) attached to analog input 0
- LED attached from digital pin 9 to ground

created 29 Oct 2008
by David A Mellis
modified 30 Aug 2011
by Tom Igoe

This example code is in the public domain.

http://www.arduino.cc/en/Tutorial/Calibration
 */

// These constants won't change:
const int sensorPin = A0;    // pin that the sensor is attached to
const int ledPin = 9;        // pin that the LED is attached to

// variables:
int sensorValue = 0;         // the sensor value
int sensorMin = 1023;        // minimum sensor value
int sensorMax = 0;           // maximum sensor value


void setup() {
  // turn on LED to signal the start of the calibration period:
  pinMode(13, OUTPUT);
  digitalWrite(13, HIGH);
```

```
  // calibrate during the first five seconds
  while (millis() < 5000) {
    sensorValue = analogRead(sensorPin);

    // record the maximum sensor value
    if (sensorValue > sensorMax) {
      sensorMax = sensorValue;
    }

    // record the minimum sensor value
    if (sensorValue < sensorMin) {
      sensorMin = sensorValue;
    }
  }

  // signal the end of the calibration period
  digitalWrite(13, LOW);
}

void loop() {
  // read the sensor:
  sensorValue = analogRead(sensorPin);

  // apply the calibration to the sensor reading
  sensorValue = map(sensorValue, sensorMin, sensorMax, 0, 255);

  // in case the sensor value is outside the range seen during calibration
  sensorValue = constrain(sensorValue, 0, 255);

  // fade the LED using the calibrated value:
  analogWrite(ledPin, sensorValue);
}
```

Upload the sketch, and let your Arduino settle with your normal ambient light levels for five seconds. Then try moving your hand over the light sensor. You should find it a lot more responsive than it is when it's just reading the analog value normally, and the LED should have a range from fully on when the light sensor is uncovered to fully off when the light sensor is covered.

Understanding the Calibration sketch

The first part of the Calibration sketch lays out all the constants and variables. The constants are the pins for the light sensor and the LED. Note that the LED should fade up and down, so it must use a PWM pin:

```
// These constants won't change:
const int sensorPin = A0;    // pin that the sensor is attached to
const int ledPin = 9;        // pin that the LED is attached to
```

The variables are used for the current sensor value and the minimum and maximum values of the sensor. You can see that `sensorMin` is initially set to a high value and `sensorMax` is set to a low one because they must work down and up, respectively, to set the minimum and maximum values:

```
// variables:
int sensorValue = 0;         // the sensor value
int sensorMin = 1023;        // minimum sensor value
int sensorMax = 0;           // maximum sensor value
```

In `setup`, quite a bit is going on. First, the usual `pinMode` sets pin 13 as an `OUTPUT`. Next is a `digitalWrite` to pin 13 to set it `HIGH`, which signals that the sensor is in its calibration phase:

```
void setup() {
  // turn on LED to signal the start of the calibration period:
  pinMode(13, OUTPUT);
  digitalWrite(13, HIGH);
```

For the first 5 seconds the sensor will calibrate. Because `millis()` starts counting the (milli)second that the program starts, the easiest way to count for 5 seconds is to use a `while` loop. The following code continues to check the value of `millis()`. As long as this value is less than 5000 (5 seconds), it carries out the code inside the curly brackets:

```
// calibrate during the first five seconds
  while (millis() < 5000) {
```

The brackets contain the calibration code. `sensorValue` stores the current sensor reading. If this reading is more than the maximum or less than the minimum (or both), the values are updated. Because this happens during five seconds, you get a number of readings, and they all help to better define the expected range:

```
sensorValue = analogRead(sensorPin);

    // record the maximum sensor value
    if (sensorValue > sensorMax) {
      sensorMax = sensorValue;
    }

    // record the minimum sensor value
    if (sensorValue < sensorMin) {
```

```
      sensorMin = sensorValue;
    }
  }
```

The LED pin is then written `LOW` to indicate that the calibration phase is over:

```
  // signal the end of the calibration period
  digitalWrite(13, LOW);
}
```

Now that the range is known, it just needs to be applied to the output LED. A reading is taken from `sensorPin`. Because the reading is in the range 0 to 1023, it must be mapped to the LED's range of 0 to 255. `sensorValue` is converted to this new range by using the `map()` function, which uses the `sensorMin` and `sensorMax` values from the calibration to specify the current range rather than using the full range of 0 to 1023:

```
void loop() {
  // read the sensor:
  sensorValue = analogRead(sensorPin);

  // apply the calibration to the sensor reading
  sensorValue = map(sensorValue, sensorMin, sensorMax, 0, 255);
```

It is still possible for the sensor to read values outside those of the calibration, so `sensorValue` must be restricted by using the `constrain ()` function. Any values outside 0 to 255 are ignored. The calibration gives a good idea of the range of values, so any larger or smaller values are likely to be anomalies:

```
  // in case the sensor value is outside the range seen during calibration
  sensorValue = constrain(sensorValue, 0, 255);
```

All that is left to do is to update the LED with the mapped and constrained value by using the `analogWrite` function on `ledPin`:

```
// fade the LED using the calibrated value:
  analogWrite(ledPin, sensorValue);
}
```

This code should give you a better representation of your sensor's changing values relative to your environment. The calibration runs only once when the program is started, so if the range still seems off, it's best to restart it or calibrate over a longer period. Calibration is designed to remove *noise* — erratic variations in the readings — so you should also make sure that the environment being measured doesn't have anything in it that you don't want to measure.

IN THIS CHAPTER

» Learning about sensors

» Understanding the complexities of different inputs

» Paying the right amount

» Knowing where to use sensors

» Wiring some examples

Chapter 11

Common Sense with Common Sensors

In my experience of teaching, I often find that when people first have an idea, they get caught up in how to carry it out using a specific piece of hardware they've found instead of focusing on what they want to achieve. If Arduino is a toolbox with the potential of solving numerous problems, using the right tool for the right job is key.

If you go to any Arduino-related site, you're likely to see a list of sensors, and finding the right ones for your project can be a baffling. A common next step is to search the web for projects similar to the one you want to do to see what other people have built. Their efforts and successes can be a great source of inspiration and knowledge, but they can also plunge you into a black hole of too many possible solutions as well as solutions that are overkill for your needs.

In this chapter, you discover not only more about different sensors and how to use them but also — and more important — *why* to use them.

Note that all prices are approximate for buying an individual sensor to give you a rough idea of the cost. If you buy in bulk or do some thorough shopping around, you should be able to make considerable savings. And to find some places to start shopping, read Chapter 16.

Making Buttons Easier

The first sensor described in this book (in Chapter 6), and arguably the best, is the pushbutton. Many kinds of pushbuttons are available. Note that switches are also included in this category. Generally, switches stick in their position in the same way that a light switch does, whereas buttons pop back. Some exceptions to this general rule are microswitches and toggle buttons. They are essentially the same electrically, and only differ mechanically.

If you plan to use a button for your project, run through the following considerations:

- **Complexity:** In its simplest form, a pushbutton can be two metal contacts that are pushed together. At its most complex, it can be a set of carefully engineered contacts in an enclosed pushbutton. Pushbuttons tend to be mounted in enclosures designed for different uses. A pushbutton like the one in your kit is perfect for breadboard layouts and suited to prototyping. If it were used in the real world, it would need protecting. Take apart an old game console controller and you may well find an enclosed pushbutton inside. If you need a more industrial button, such as an emergency stop button, the switch may be larger and more robust, and may even contain a bigger spring to handle the force of someone hitting or kicking it.

  TIP

  The great thing about pushbuttons is that they never get complicated. However, the right spring or click to a button can make all the difference in the quality of your project, so choose wisely.

- **Cost:** The cost of a pushbutton varies greatly depending on the quality of the enclosure and the materials used. The prices for RS Components range from 10 cents for a microswitch to around $150 for an industrial stop button in an enclosure. It's possible to use cheaper buttons for most applications.

- **Location:** You can use buttons to detect presses from intentional human contact (or even unintentional, if you are clever with how you house the buttons). Museum exhibits are a great example of using buttons to register intentional human contact. People "get" how to use buttons. They're everywhere, and people use them every day without thinking. Sometimes it may seem clever to use subtler methods, but if in doubt, a button is always a safe option.

  Also consider how you might apply the use of a button to what's already in place. For example, maybe you're monitoring how often a door is opened in your house. If you put a highly sensitive microswitch against the door when it's closed, that switch tells you every time the door moves away from it.

In Chapters 6 and 10, you learn how to wire a button circuit and how to refine it, respectively. In the example in the following section, you learn how to simplify the hardware of your button. By using a hidden feature of your Arduino, you can use a button with no additional hardware.

Implementing the DigitalInputPullup sketch

The basic button circuit is a simple one, but it can be made simpler still by using a little-known function on your microcontroller. On the basic button example in Chapter 6, a *pull-down* resistor is connected to ground to make the button pin read `LOW`. Whenever pressed, the button is connects to 5V and goes `HIGH`. This behavior allows you to read the button's state as an input.

In the microcontroller, an internal *pull-up* resistor can be activated to give you a constant `HIGH` value. When a button connected to ground is pressed, it grounds the current and sets the pin to `LOW`. This design gives you the same functionality as the basic example from Chapter 6, but the logic is inverted: `HIGH` is an open switch, and `LOW` is a closed switch. The wiring is, therefore, simpler because you eliminate the need for an extra wire and resistor.

To complete this example, you need the following:

- An Arduino Uno
- A breadboard
- A pushbutton
- An LED (optional)
- Jump wires

Complete the circuit shown in Figures 11-1 and 11-2 to try the new, simpler pushbutton using the DigitalInputPullup sketch.

REMEMBER

An LED on the board is already linked to pin 13, but if you want to accentuate the output, you can insert an LED straight into pin 13 and its neighboring GND pin.

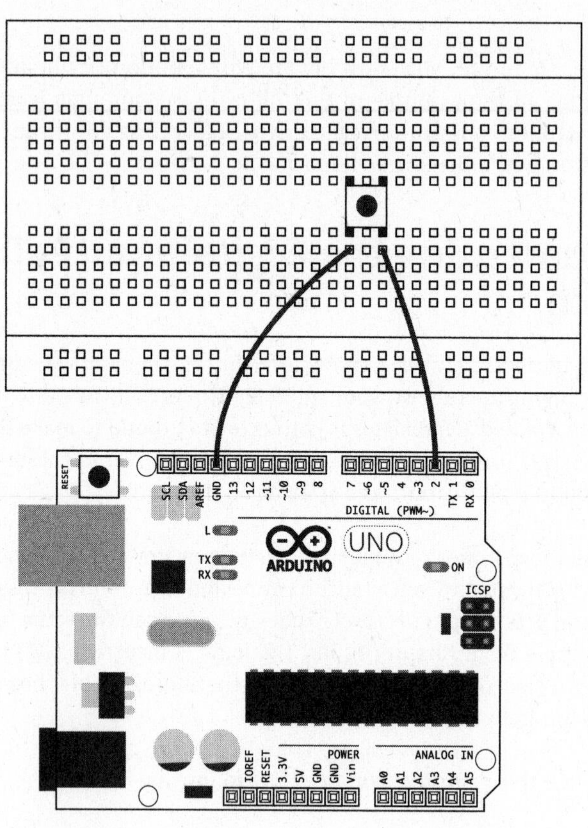

FIGURE 11-1: The pushbutton circuit layout.

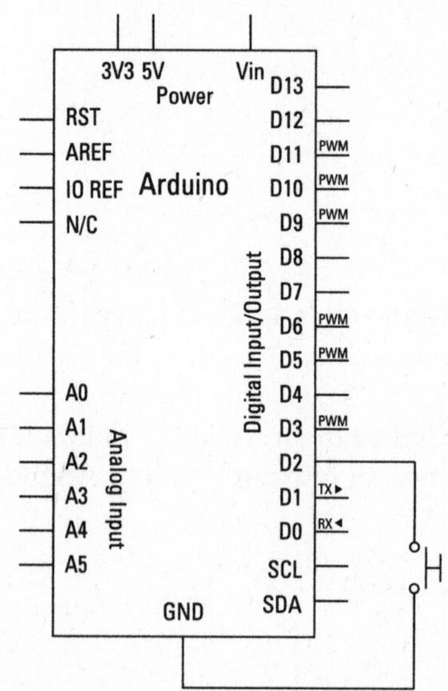

FIGURE 11-2: A circuit diagram of the pushbutton circuit.

Complete the circuit and choose File ➪ Examples ➪ 02.Digital ➪ DigitalInputPullup to find the DigitalInputPullup sketch.

```
/*

Input Pull-up Serial

This example demonstrates the use of pinMode(INPUT_PULLUP). It reads a
Digital input on pin 2 and prints the results to the Serial Monitor.

The circuit:
- momentary switch attached from pin 2 to ground
- built-in LED on pin 13

Unlike pinMode(INPUT), there is no pull-down resistor necessary. An internal
20K-ohm resistor is pulled to 5V. This configuration causes the input to
read HIGH when the switch is open, and LOW when it is closed.

created 14 Mar 2012
by Scott Fitzgerald

This example code is in the public domain.

http://www.arduino.cc/en/Tutorial/InputPullupSerial
*/

void setup(){
  //start serial connection
  Serial.begin(9600);
  //configure pin2 as an input and enable the internal pull-up resistor
  pinMode(2, INPUT_PULLUP);
  pinMode(13, OUTPUT);

}

void loop(){
  //read the pushbutton value into a variable
  int sensorVal = digitalRead(2);
  //print out the value of the pushbutton
  Serial.println(sensorVal);

  // Keep in mind the pullup means the pushbutton's
  // logic is inverted. It goes HIGH when it's open,
```

```
  // and LOW when it's pressed. Turn on pin 13 when the
  // button's pressed, and off when it's not:
  if (sensorVal == HIGH) {
    digitalWrite(13, LOW);
  }
  else {
    digitalWrite(13, HIGH);
  }
}
```

Understanding the DigitalInputPullup sketch

The DigitalInputPullup sketch is similar to the standard button sketch but with a few changes. In `setup`, serial communication is started to monitor the state of the button. Next, the `pinMode` of the inputs and outputs is set. Pin 2 is your button pin, but instead of setting it to `INPUT`, you use `INPUT_PULLUP`. Doing so activates the internal pull-up resistor. Pin 13 is set to be an output as an LED control pin:

```
void setup(){
  //start serial connection
  Serial.begin(9600);
  //configure pin2 as an input and enable the internal pull-up resistor
  pinMode(2, INPUT_PULLUP);
  pinMode(13, OUTPUT);

}
```

In the main loop, you read the value of the pull-up pin and store it in the `sensorVal` variable. You then print the variable to the serial monitor to show you what value is being read:

```
void loop(){
  //read the pushbutton value into a variable
  int sensorVal = digitalRead(2);
  //print out the value of the pushbutton
  Serial.println(sensorVal);
```

But because the logic is inverted, you need to invert your `if ()` statement to make it correct. A `HIGH` value is open and a `LOW` value is closed. Inside the `if ()` statement, you can write any actions to perform. In this case, the LED is being turned off, or set `LOW`, whenever the button pin is open, or pulled `HIGH`:

```
  // Keep in mind the pullup means the pushbutton's
  // logic is inverted. It goes HIGH when it's open,
  // and LOW when it's pressed. Turn on pin 13 when the
  // button's pressed, and off when it's not:
  if (sensorVal == HIGH) {
    digitalWrite(13, LOW);
  }
  else {
    digitalWrite(13, HIGH);
  }
}
```

This method is great when you don't have enough spare components, because it allows you to make a switch with only a couple of wires, if necessary. This functionality can be used on any digital pin, but only for inputs.

Exploring Piezo Sensors

In Chapter 7, you learn how to make sound using a piezo buzzer, but you should know that you have another way to use the same hardware as an input rather than as an output. To make a sound with a piezo, you put a current through it and it vibrates, so it follows that if you vibrate the same piezo, you generate a small amount of electrical current. Using a piezo in this way allows you to create a *knock sensor* and is used to measure vibrations on the surface to which it is fixed.

Piezos vary in size, and that determines the scale of the vibrations that they can detect. Small piezos are extremely sensitive to vibrations and need little to max out their range. Bigger piezos have a broader range, but more vibration is necessary to register a reading. In addition, specialized piezo sensors can act as inputs to detect flex, touch, vibration, and shock. These cost slightly more than a basic piezo element, and are usually made from a flexible film, which makes them a lot more robust.

When using a piezo, consider the following:

- **Complexity:** Piezos are relatively simple to wire, needing only a resistor to function in a circuit. Because the top half is made of a fragile ceramic, it is often enclosed in a plastic case, which makes it easy to mount and avoids any direct contact with the fragile solder joints on the surface of the piezo.

- **Cost:** Piezo elements are inexpensive, costing from around 20 cents for the cheapest elements without a casing to $15 for high-power piezo buzzers. As an input, a piezo element is preferable to the more specific piezo buzzer. The usual difference is a smaller form factor for buzzers, whereas elements usually have a broader base. The latter is preferable for the knock sensor because it gives you more area on the piezo as well as more contact with the surface being monitored.

  Piezos are much cheaper to purchase from the major electronics companies, but because these require you to browse their vast online catalogues, you might find it more useful to buy a selection from retail stores first, such as RadioShack, where you can see the product in real life and get a feel for the different shapes, styles, and housings.
- **Location:** Knock sensors are not usually used as a direct input. Because they are so fragile, having people tapping on them all the time is risky. Instead, fix your piezo to a rigid surface such as wood, plastic, or metal, and let that surface take the punishment. For example, a knock sensor mounted on a staircase could be discreet and unobtrusive but still give highly accurate readings.

Piezos are simple and inexpensive sensors with a large variety of uses. You can use them to detect vibrations or more directly in a homemade electric drum kit. This section's example shows you how to wire your own set of piezo knock sensors.

Implementing the Knock sketch

Knock sensors use a piezo element to measure vibration. When a piezo vibrates, it produces a voltage that can be interpreted by your Arduino as an analog signal. Piezo elements are more commonly used as buzzers, which inversely make a vibration when a current is passed through them.

You need the following:

- An Arduino Uno
- A breadboard
- A piezo
- A 1M ohm resistor
- Jump wires

Using the layout and circuit diagrams in Figures 11-3 and 11-4, assemble the circuit for the knock sensor. The hardware in this circuit is similar to the piezo buzzer sketch in Chapter 7, but with a few changes you can make this piezo element into an input as well.

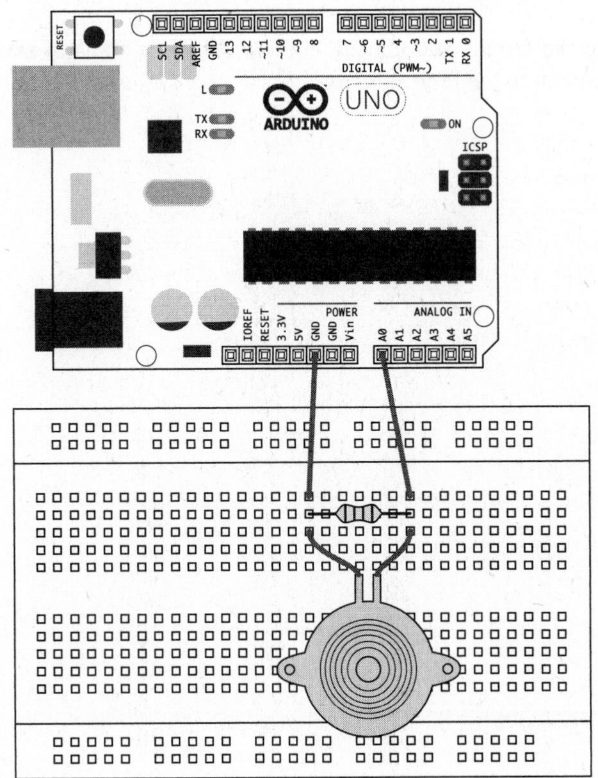

FIGURE 11-3: A knock sensor circuit layout.

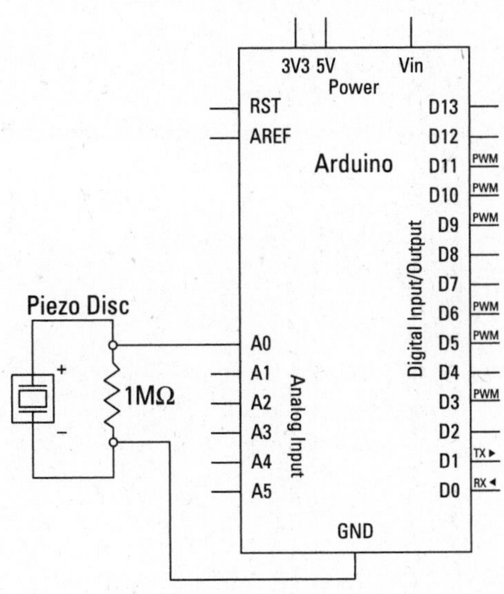

FIGURE 11-4: A knock sensor circuit diagram.

Complete the circuit and choose File ➪ Examples ➪ 06.Sensors ➪ Knock from the Arduino menu to load the sketch.

```
/*
Knock Sensor

This sketch reads a piezo element to detect a knocking sound.
It reads an analog pin and compares the result to a set threshold.
If the result is greater than the threshold, it writes "knock" to the serial
port, and toggles the LED on pin 13.

The circuit:
- positive connection of the piezo attached to analog in 0
- negative connection of the piezo attached to ground
- 1 megohm resistor attached from analog in 0 to ground

created 25 Mar 2007
by David Cuartielles <http://www.0j0.org>
modified 30 Aug 2011
by Tom Igoe

This example code is in the public domain.

http://www.arduino.cc/en/Tutorial/Knock
*/


// these constants won't change:
const int ledPin = 13;       // led connected to digital pin 13
const int knockSensor = A0; // the piezo is connected to analog pin 0
const int threshold = 100;  // threshold value to decide when the detected
                            // sound is a knock or not


// these variables will change:
int sensorReading = 0;      // variable to store the value read from the
                            // sensor pin
int ledState = LOW;         // variable used to store the last LED status,
                            // to toggle the light

void setup() {
 pinMode(ledPin, OUTPUT); // declare the ledPin as OUTPUT
 Serial.begin(9600);       // use the serial port
}
```

```
void loop() {
  // read the sensor and store it in the variable sensorReading:
  sensorReading = analogRead(knockSensor);

  // if the sensor reading is greater than the threshold:
  if (sensorReading >= threshold) {
    // toggle the status of the ledPin:
    ledState = !ledState;
    // update the LED pin itself:
    digitalWrite(ledPin, ledState);
    // send the string "Knock!" back to the computer, followed by newline
    Serial.println("Knock!");
  }
  delay(100);  // delay to avoid overloading the serial port buffer
}
```

Press the Compile button to check your code. Doing so highlights any grammatical errors and lights them up in red. If the sketch compiles correctly, click Upload to send the sketch to your board. When it is finished uploading, choose the serial monitor and give the surface that your piezo is on a good knock. If it's working, *Knock!* appears on the serial monitor and the LED changes with each successful knock.

If nothing happens, double-check your wiring:

- Make sure that you're using the correct pin number.
- Check the connections on the breadboard. If the jump wires or components are not connected using the correct rows in the breadboard, they will not work.

Understanding the Knock sketch

In the Knock sketch, the first declarations are constant values, the LED pin number, the knock sensor pin number, and the threshold for a knock value. These are set and don't change throughout the sketch:

```
// these constants won't change:
const int ledPin = 13;       // led connected to digital pin 13
const int knockSensor = A0; // the piezo is connected to analog pin 0
const int threshold = 100;  // threshold value to decide when the
                            // detected sound is a knock or not
```

Two variables do change — the current sensor reading and the state of the LED:

```
// these variables will change:
int sensorReading = 0;      // variable to store the value read from the sensor
   pin
int ledState = LOW;         // variable used to store the last LED status,
                            // to toggle the light
```

In `setup`, the LED pin is set to be an output and the serial port is opened for communication:

```
void setup() {
 pinMode(ledPin, OUTPUT); // declare the ledPin as OUTPUT
 Serial.begin(9600);      // use the serial port
}
```

The first line in the loop reads the analog value from the knock sensor pin:

```
void loop() {
  // read the sensor and store it in the variable sensorReading:
  sensorReading = analogRead(knockSensor);
```

This value is compared to the threshold value:

```
  // if the sensor reading is greater than the threshold:
  if (sensorReading >= threshold) {
```

If the value of `sensorReading` is greater than or equal to the `threshold` value, the LED's state is switched between 0 and 1 using `!`, the NOT symbol. The `!` in this case is used to return the opposite Boolean value of the `ledState` variable's current value. As you know, Booleans are either 1 or 0 (true or false), the same as the possible values of `ledState`. This line of code could be written as "make `ledState` equal to whatever value it is not:"

```
    // toggle the status of the ledPin:
    ledState = !ledState;
```

The `ledState` value is then sent to the LED pin using `digitalWrite`. The digitalWrite function interprets a value of 0 as `LOW` and 1 as `HIGH`:

```
    // update the LED pin itself:
    digitalWrite(ledPin, ledState);
```

Finally, *Knock!* is sent to the serial port with a short delay, for stability:

```
// send the string "Knock!" back to the computer, followed by newline
    Serial.println("Knock!");
  }
  delay(100);  // delay to avoid overloading the serial port buffer
}
```

Utilizing Pressure, Force, and Load Sensors

Three closely related kinds of sensors are commonly confused: pressure, force, and load sensors. These three sensors are extremely different in how they behave and what data they can give you, so it's important to know the difference so that you choose the one that's right for your situation. In this section, you learn about the different definitions of each of these sensors, as well as where you use them and why you use one over the other.

Consider the following as you plan:

- **Complexity:** As you might expect, complexity increases depending on how accurate you need to be:
  - *Pressure pads* are designed to detect when pressure is applied to an area, and they come in a variety of both quality and accuracy. The simplest pressure pads are often misnamed and are really the equivalent of big switches. Inside a simple pressure pad are two layers of foil separated by a layer of foam with holes in it. When the foam is squashed, the metal contacts touch through the foam and complete the circuit. Instead of measuring pressure or weight, the pad is detecting when there is enough weight to squash the foam. These pads do a fine job and are similar to the mechanisms found inside dance mats — ample proof that you don't need to overthink your sensors!
  - For more precision, you may want to use *force sensors*, which measure the force applied by whatever is put on them within their range. Although force sensors are accurate enough to detect a change in weight, they are not accurate enough to provide a precise measurement. Force sensors are usually flexible, force-sensitive resistors — that is, resistors made on a flexible PCB that change their resistance when force is applied. The resistor itself is on the flexible circuit board, and although the board can tolerate extremely high forces and loads, protecting it from direct contact is a good idea to prevent it from bending, folding, or tearing.

- With pressure pads at one end of the spectrum, at the other are load sensors. An example of a load sensor is found in your bathroom scale. *Load sensors* can accurately measure weight up to their limit. They work in much the same way as force sensors by changing resistance as they bend. In most cases, a load sensor is fixed to a rigid piece of metal and monitors changes as the metal is put under strain. The changes are so minute that they often require an amplification circuit known as a Wheatstone bridge. Incorporating this kind of sensor is more complex than the others, but you can find material on the Internet that will walk you thorough the process.

» **Cost:** The cost of each sensor is relatively low, even for the most sensitive ones. All the materials to make a cheap pressure pad will set you back $3; the cost for an inexpensive, entry-level pressure mat available from most electronics stores and suppliers is $12. Force-sensitive resistors range from $8 to $23, but cover a much smaller area than a pressure pad, so you may need quite a few of them to cover a large area. At around $11, load sensors are also relatively cheap, most likely because they are so widespread that mass production has knocked the price down. There may be an extra cost in time to plan and make the additional circuitry.

» **Location:** The challenge with all these sensors is housing them to prevent damage. In the case of the pressure pad and force-sensitive resistors, placing a thick layer of upholstery foam on the side that the force is coming from is a good idea. Depending on the density of the foam, it should dampen enough of the force to protect the sensor but still compress it enough for a good reading. Underneath the sensors, you want to have a solid base to give you something to push against. This base could be the floor or surface that the sensor is placed on, or you could attach a piece of MDF/plywood to the underside of the sensor. It's a good idea to protect the exterior of your pressure sensor as well, so consider something a bit sturdier than foam on the exterior. For a soft finish, upholstery vinyl is a great option. If you plan to have people walk on the surface, for example, a layer of wood on the top to sandwich the foam will spread the load and can easily be replaced, if needed.

  Load sensors require little movement and should be connected to or placed in direct contact with a ridged surface. If you use a load sensor in a bathroom scale, it may take some trial and error to place the sensor in the correct location to get an accurate reading. Sometimes multiple sensors are used to get an average reading across the surface.

After you choose a sensor, you need to figure out how to use it:

» *Pressure pads* are an extremely simple circuit, the same as for a pushbutton. The hardware of a pressure pad is also easy enough that you can make one yourself using two sheets of foil (or the more flexible conductive fabric or conductive thread), a sheet of foam, a cover, and a couple of wires.

- *Force sensors* are also relatively easy to use and can take the place of other analog sensors, such as light or temperature sensors, in simple Arduino circuits. The ranges of force may vary, but whatever the range, scaling the force to your needs in the code is simple.
- *Load sensors* are probably the most complex sensor if they're being used for accurate reading, as with a set of weight scales. They require extra circuitry and an amplifier for the Arduino to read the minute changes in resistance. This topic is outside the scope of this book, so if you want to know more, get friendly with Google.

Force sensors are just like any other variable resistor and can easily be switched with potentiometer or light-dependent resistors as needed. In this section's example, you use force-sensitive resistors and the toneKeyboard sketch to make an Arduino piano keyboard.

Implementing the toneKeyboard sketch

You may think of pushbuttons as the perfect input for a keyboard, but force-sensitive resistors give you much more touch sensitivity. Rather than detect just a press, you can detect also the intensity of the keypress, in the same way as on a traditional piano.

You need the following:

- An Arduino Uno
- A breadboard
- Three force-sensitive resistors
- Three 10k ohm resistors
- One 100 ohm resistor
- A piezo element
- Jump wires

Using the layout and circuit diagrams in Figures 11-5 and 11-6, lay out the force-sensitive resistors and the piezo to make your own keyboard.

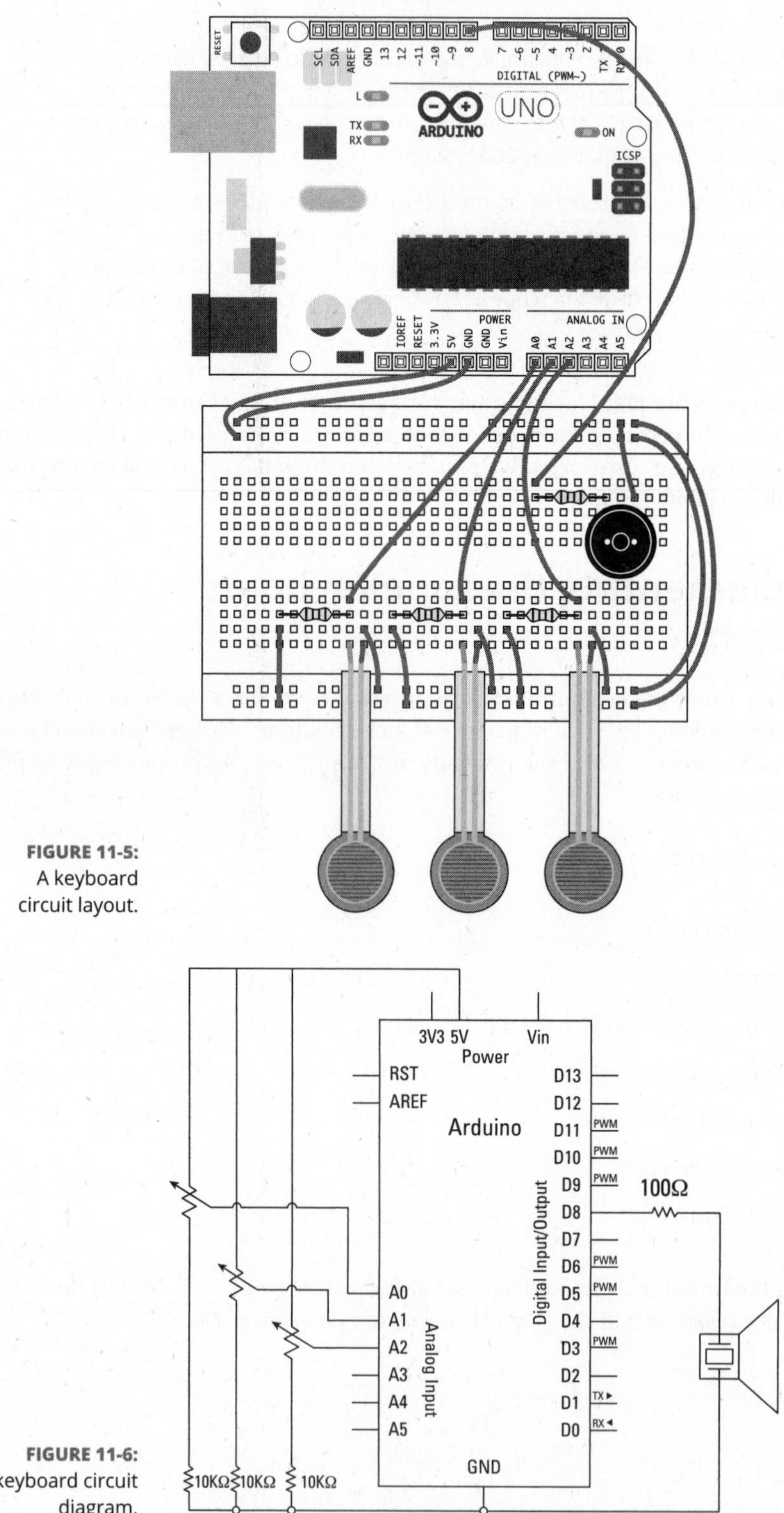

FIGURE 11-5: A keyboard circuit layout.

FIGURE 11-6: A keyboard circuit diagram.

Complete the circuit and choose File ➪ Examples ➪ 02.Digital ➪ toneKeyboard from the Arduino menu to load the sketch.

```
/*
  Keyboard

  Plays a pitch that changes based on a changing analog input

  circuit:
  - three force-sensing resistors from +5V to analog in 0 through 5
  - three 10 kilohm resistors from analog in 0 through 5 to ground
  - 8 ohm speaker on digital pin 8

  created 21 Jan 2010
  modified 9 Apr 2012
  by Tom Igoe

  This example code is in the public domain.

  http://www.arduino.cc/en/Tutorial/Tone3
*/

#include "pitches.h"

const int threshold = 10;    // minimum reading of the sensors that
                             // generates a note

// notes to play, corresponding to the 3 sensors:
int notes[] = {
 NOTE_A4, NOTE_B4,NOTE_C3
};

void setup() {

}

void loop() {
 for (int thisSensor = 0; thisSensor < 3; thisSensor++) {
  // get a sensor reading:
  int sensorReading = analogRead(thisSensor);

  // if the sensor is pressed hard enough:
  if (sensorReading > threshold) {
```

```
    // play the note corresponding to this sensor:
    tone(8, notes[thisSensor], 20);
   }
  }
}
```

Press the Compile button to check your code. The compiler should highlight any grammatical errors and light them up in red. If the sketch compiles correctly, click Upload to send the sketch to your board. When it has finished uploading, try the keys to make sure they're working. If they are, you're ready to play.

If nothing happens, double-check your wiring:

- Make sure that you're using the correct pin number.
- Check the connections on the breadboard. If the jump wires or components are not connected using the correct rows in the breadboard, they won't work.

Understanding the toneKeyboard sketch

The toneKeyboard sketch uses the same table of notes as the Melody sketch in Chapter 7. The first line includes `pitches.h`, which should be open in a separate tab next to the main sketch:

```
#include "pitches.h"
```

A low threshold of 10 (out of a possible 1024) is set to avoid any low readings resulting from background vibrations:

```
const int threshold = 10;  // minimum reading of the sensors that
                           // generates a note
```

The notes for each sensor are stored in an array with values (0, 1, and 2) that correspond to the analog input pin numbers (A0, A1, and A2). You can change the note values in the array manually by using the look-up table on `pitches.h`. Simply copy and paste new note values in the change the note of each sensor.

```
// notes to play, corresponding to the 3 sensors:
int notes[] = {
 NOTE_A4, NOTE_B4,NOTE_C3
};
```

In setup, you have nothing to define because the analog input pins are set to be inputs by default:

```
void setup() {

}
```

In the main loop, a for () loop cycles through the numbers 0 to 2:

```
void loop() {
 for (int thisSensor = 0; thisSensor < 3; thisSensor++) {
```

The value of the for () loop, which is used as the pin number, is stored temporarily to sensorReading:

```
  // get a sensor reading:
  int sensorReading = analogRead(thisSensor);
```

If the reading is greater than the threshold, it is used to trigger the correct note assigned to that input:

```
  // if the sensor is pressed hard enough:
  if (sensorReading > threshold) {
   // play the note corresponding to this sensor:
   tone(8, notes[thisSensor], 20);
  }
 }
}
```

Because the loop happens so quickly, any delay in reading each sensor is unnoticeable.

Sensing with Style

Capacitive sensors detect changes in electromagnetic fields. Every living thing — even you — has an electromagnetic field. Capacitive sensors are extremely useful because they can detect human contact and ignore other environmental factors. You're probably familiar with high-end capacitive sensors because they are present in nearly all smartphones, but they have been around since the late 1920s. You can find Arduino kits with capacitive sensors that you can hook up easily, such as capacitive touch keypads. But it's just as easy to make your own capacitive sensors with an Arduino and an antenna.

Consider the following in your plans:

- **Complexity:** Because all that is required is an antenna, you can be creative with what the antenna is and where it is placed. Short pieces of wire or copper tape are great for simple touch sensors. The piece of copper tape suddenly becomes a touch switch, meaning that you don't even need a pushbutton to get the same functionality. You could even connect the antenna to a bigger metal object such as a lamp, turning it into a touch lamp.

  If the antenna is made from a reel of wire or a piece of foil, you can extend the range of the sensor beyond touch, which is known as a *projected capacitive sensor.* Using a capacitive sensor you can detect a person's hand a few inches away from the antenna, which creates a lot of new possibilities for hiding sensors behind other materials. These discreet capacitive sensors are now commonly seen in many recent consumer electronics to remove physical buttons and maintain the sleek shape of the product. The electronics can also be placed under layers of other material, protected from the outside world.

  Capacitive touch sensors are easy to make. The difficulty with projected field sensors to determine the range of the field. The best way to determine this range is by experimentation, testing to see whether the field that you're generating is far-reaching enough.

- **Cost:** A capacitive touch kit designed for a specific purpose costs around $15 to $25. The kit should perform its job well, but it will be limited to the design of the interface. A capacitive sensor breakout board from SparkFun for around $10 lets you control up to 12 capacitive sensors. You have to wire your own touchpads, but you're free to design an interface that suits your purpose.

  The cheapest option is to use the CapSense library for Arduino, which allows you to make a capacitive sensor with an antenna and no additional hardware! You can spend a few cents for an antenna or repurpose an old one.

- **Location:** Capacitive touch sensors can work with any conductive metal, so if you can design an attractive metal exterior, the only work will be to connect that exterior to your Arduino. If you're looking for something more discreet, you may want to experiment with different layers of wood or plastic to hide your metal antenna. A thin layer of plywood allows the metal to be close to the surface, able to trigger the sensor. By covering the antenna with a non-conductive surface, you also give it a seemingly magical property, ensuring that people are left guessing at how it works.

The easiest way to make a capacitive sensor is to use the CapSense library by Paul Badger. By using the CapSense library (I explain how in the "Getting the CapSense Library" sidebar), you can replace the mechanical switches with highly robust capacitive touch sensors or capacitive presence detectors.

GETTING THE CapSense LIBRARY

The CapSense library is available from GitHub, an online repository of software that manages different versions and allows you to see who has updated the software, and how. It's an excellent system for sharing and collaborating on code projects. You can find the Arduino platform on GitHub; check it out if you're curious about any changes. To get the library:

1. **Point your web browser to the GitHub CapSense page at** `github.com/moderndevice/CapSense`.
2. **On the CapSense page, click the Clone button or download and then click Download ZIP, as shown in the figure.**

   The latest version of the library is downloaded to your downloads folder or a folder you specify.

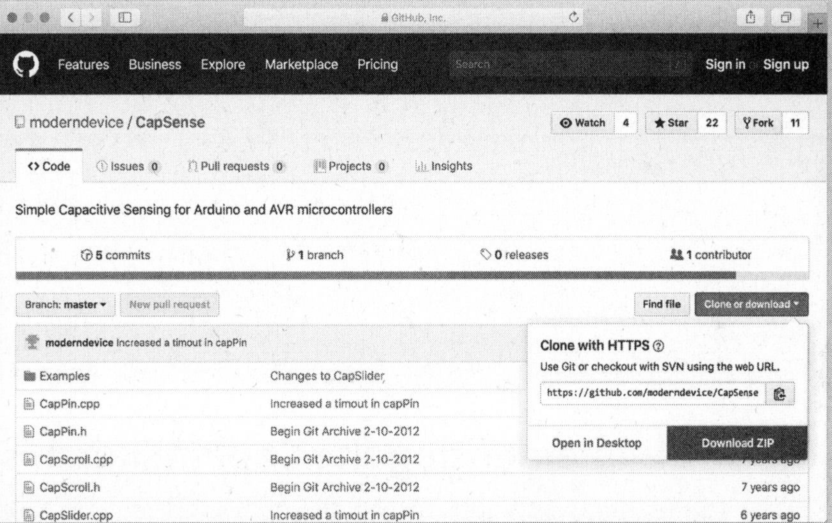

3. **Rename the folder CapSense.**

   Inside the folder, you should see a number of files ending in `.h` and `.cpp` as well as an Examples folder.
4. **Move the entire folder to your Arduino libraries directory.**

   This directory should be the same as the one that your sketches are saved to, for example: Username/Documents/Arduino/libraries. If you don't have a libraries directory, create one.

(continued)

Your can find your Arduino Save directory by choosing Arduino ⇨ Preferences from the Arduino menu bar. After the CapSense library is inside this folder, it will be available the next time you run Arduino.

5. **Start or restart Arduino and go to Sketch ⇨ Include Library in the Arduino menu.**

   Look for CapSense under the Contributed libraries section. If you don't find it, check your directories and spelling and then restart Arduino.

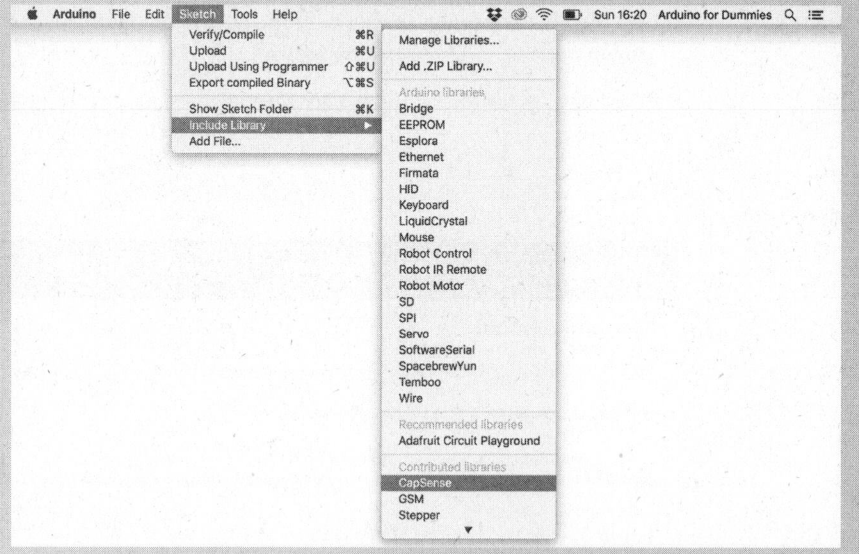

Implementing the CapPinSketch sketch

For this project, you need the following:

- » An Arduino Uno
- » A wire antenna
- » Crocodile clips (optional)

As you can see from the photo in Figure 11-7, little work is needed. You can simply have a wire antenna connected to pin 5, and you can enlarge your antenna by connecting it to any other conductive surface. Crocodile clips are useful to latch onto different antennas quickly and easily.

FIGURE 11-7: A photo of a DIY capacitive sensor.

Build the CapSense circuit, and choose File ➪ Examples ➪ CapSense ➪ Examples ➪ CapPinSketch from the Arduino menu to load the sketch.

```
#include <CapPin.h>

/* CapPin
 * Capacitive Library CapPin Demo Sketch
 * Paul Badger 2011
 * This class uses the built-in pullup resistors read the capacitance
 * on a pin.
 * The pin is set to input and then the pullup is set,
 * A loop times how long the pin takes to go HIGH.
 * The readPin method is fast and can be read 1000 times in under 10 mS.
 * By reading the pin repeated you can sense "hand pressure"
 * at close range with a small sensor. A larger sensor (piece of foil/metal)
 * will yield
 * larger return values and be able to sense at more distance. For
 * a more sensitive method of sensing pins see CapTouch
 * Hook up a wire with or without a piece of foil attached to the pin.
 * I suggest covering the sensor with mylar, packing tape, paper or other
 * insulator to avoid having users directly touch the pin.
 */
```

```
CapPin cPin_5 = CapPin(5);    // read pin 5

float smoothed;

void setup() {

  Serial.begin(115200);
  Serial.println("start");
  // slider_2_7.calibrateSlider();

}

void loop() {

  delay(1);
  long total1 = 0;
  long start = millis();
  long total =  cPin_5.readPin(2000);

  // simple lowpass filter to take out some of the jitter
  // change parameter (0 is min, .99 is max) or eliminate to suit
  smoothed = smooth(total, .8, smoothed);

Serial.print( millis() - start);      // time to execute in mS
Serial.print("\t");
Serial.print(total);                  // raw total
Serial.print("\t");
Serial.println((int) smoothed);       // smoothed
delay(5);
}

// simple lowpass filter
// requires recycling the output in the "smoothedVal" param
int smooth(int data, float filterVal, float smoothedVal){

  if (filterVal > 1){      // check to make sure param's are within range
    filterVal = .999999;
  }
  else if (filterVal <= 0){
    filterVal = 0;
  }
```

```
  smoothedVal = (data * (1 - filterVal)) + (smoothedVal * filterVal);

  return (int)smoothedVal;
}
```

Press the Compile button to check your code. Compiling highlights any grammatical errors, which light up in red. If the sketch compiles correctly, click Upload to send the sketch to your board. When it has finished uploading, open the serial monitor, set the baud rate to 115200, and touch or approach the antenna. You should see two values racing down the screen. On the left is the raw value being read; on the right is the same reading after smoothing.

If nothing happens, double-check your wiring:

- Make sure that you're using the correct pin number.
- Check the connections on the breadboard. If the jump wires or components are not connected using the correct rows in the breadboard, they will not work.

Understanding the CapPinSketch sketch

At the start of the CapPinSketch sketch in the declarations, a new `CapPin` object is named. Note that `cPin_5` is the name and it is assigned to pin 5 using `CapPin(5)`:

```
CapPin cPin_5 = CapPin(5);    // read pin 5
```

A float named `smoothed` is declared to store the processed value of the sensor:

```
float smoothed;
```

WHAT IS A FLOAT?

A *float,* or floating-point number, is any number with a decimal point. Variables can be set to floating-point numbers instead of integers. This is preferable in some situations, such as when you're taking extremely precise readings of capacitance. However, floats take much more time to process than integers and, therefore, should be avoided if possible.

In `setup`, serial communication is started on a baud rate of 115200, and the message *start* is sent to indicate to you that the serial port is connected:

```
void setup()
{

  Serial.begin(115200);
  Serial.println("start");
```

This commented line is not used in this sketch but is referenced in some of the other CapSense examples. It can be uncommented to include further calibration functions that are in the library but is not be covered in this example:

```
// slider_2_7.calibrateSlider();

}
```

In this sketch, many variables are declared locally. Because they're not needed outside the loop, they're removed after each loop and redeclared at the start of the next loop.

First, a one-millisecond delay occurs to help improve the stability of the reading:

```
void loop()
{

  delay(1);
```

Next, the long variable `total1` is declared. This variable can look confusing because the lowercase *L* and the numeral 1 look the same in most fonts. Incidentally, this variable is not used in this sketch. It may well be left over from a previous version:

```
  long total1 = 0;
```

The next long variable is set to the current `millis ()` value. Because this variable is local, the value is reset on each loop:

```
  long start = millis();
```

The specific function `.readPin()` reads your capacitive pin:

```
  long total =  cPin_5.readPin(2000);
```

If you want to explore in more depth what's happening, look at `CapPin.cpp` in the CapSense library. At first, the CapPin.cpp file looks baffling, but by reading the line that follows, you can see that the value relates to the number of samples that the Arduino is taking of the capacitance reading:

```
long CapPin::readPin(unsigned int samples)
```

TIP

Editing the inner workings of libraries is not advised for beginners, but it is good to have a look at them to know what's happening in your code and try to gain a better understanding of them.

A smoothing function is also included in the sketch. This function takes the raw reading from the sensor, the smoothing value, and then the output variable. At present, the smoothing value is set to 0.8, but go ahead and experiment with this value to find the appropriate amount of smoothing for your application. This amount is dependent on how fast the loop is completed and how many readings are made in that time, so bear that in mind if you expect to add a lot of other controls or outputs:

```
// simple lowpass filter to take out some of the jitter
  // change parameter (0 is min, .99 is max) or eliminate to suit
  smoothed = smooth(total, .8, smoothed);
```

Finally, the values are printed to the serial port to be monitored. `millis() - start` gives the time that is taken to carry out the reading. If more samples are taken or any delays are added to the code, these activities increase the time to complete the loop and, therefore, the reaction time of the sensor:

```
Serial.print( millis() - start);      // time to execute in mS
```

Tabs are used to neatly space the values. The total and smoothed values are both printed for comparison. You may notice a slight delay in the response time of the smoothed value. This delay shows you that your Arduino is reading many more values to do the smoothing, which takes time. This delay is barely noticeable when the sensor is in use because the baud rate is so high:

```
Serial.print("\t");
Serial.print(total);                 // raw total
Serial.print("\t");
Serial.println((int) smoothed);      // smoothed
  delay(5);
}
```

At the bottom of the sketch outside the main loop is an additional function, referred to as a *lowpass filter*, which gives you the smoothed result. You can see that rather than starting with `void`, as is the case in `setup ()` and `loop ()`, the function starts with `int`, which means that an integer value is returned. Starting with `int` indicates that this function returns an integer value rather than a floating-point value.

```
// simple lowpass filter
// requires recycling the output in the "smoothedVal" param
int smooth(int data, float filterVal, float smoothedVal){

  if (filterVal > 1){       // check to make sure param's are within range
    filterVal = .999999;
  }
  else if (filterVal <= 0){
    filterVal = 0;
  }

  smoothedVal = (data * (1 - filterVal)) + (smoothedVal  *  filterVal);

  return (int)smoothedVal;
}
```

Tripping Along with Lasers

Laser trip wires are made up of two parts, a laser light source and a light sensor. As you know from the movies, when the beam is broken, an alarm sounds and the henchmen come running. With an Arduino, you can make a laser trip wire simply and trigger anything you want from it. Rather than buy a state-of-the-art security system, you can build one yourself using a few simple components. You're on your own for the henchmen, though.

- **Complexity:** Lasers are a difficult subject area because of the potential risks of working with them. But rather than risk your eyesight or spend a few years studying, why not use something that's already been tested, certified, and turned into a product? Laser pens or laser pointers are widely available and relatively cheap. These are usually Class 1 lasers, the lowest class, and are visible and safe under all conditions of normal use, but you're still advised to check the specifications of your laser to ensure that it's appropriate for the audience and the environment. Adults are normally sensible enough not to look directly into the laser, but with children, you want to err on the side of caution and find another sensor.

  Because a laser beam is so precise, it's best to choose a large sensor to detect it so that you have a large area to aim for. The only added complexity of this sensor may be powering the laser. A laser pointer is usually battery powered (because it's not much use with a huge plug on the end), so you may need to replace the batteries every few days or wire a power supply of equal value to the battery compartment.

To make the laser and the light sensor useful in the outside world, I recommend that you house them both in enclosures. A nice touch is to mount the enclosures on mini tripods to give you some flexibility when aligning them.

- **Cost:** For around $15 you can purchase a small, discreet laser pointer from RadioShack. Battery life and beam color are the main differences between pointers. The light sensor costs around $0.75 to $2.50, depending on the size. If you opt for an enclosure, it may cost around $6. A mini tripod is around $10.
- **Location:** If you have a fixed location for your trip wire, mounting the enclosure on either side of a doorway is simple. The best position is as low to the floor as possible to avoid eye contact with the laser. If you're not sure where you want the trip wire or want to try a few ideas, keep it mobile and take your tripod-mounted trip wire with you wherever it's needed. Laser trip wires make especially good camera triggers, which you can read more about in the "Hacking Other Hardware" bonus chapter at `www.dummies.com/go/arduinofd`.

The laser trip wire is a refinement of a conventional light sensor. By providing a more intense, controlled light source, you can increase the accuracy of a simple light sensor.

In this example, you use a laser to make your light sensor achieve its full potential. By using the AnalogInOutSerial circuit, you can monitor the levels of your sensor when the laser is hitting it and look for the change when the laser is obscured. From this reading, you can trigger any number of outputs.

Implementing the AnalogInOutSerial sketch

To implement the AnalogInOutSerial sketch, you need the following:

- An Arduino Uno
- A breadboard
- A light sensor
- An LED
- A 10k ohm resistor
- A 220 ohm resistor
- Jump wires

Complete the circuit shown in Figures 11-8 and 11-9 to make the receiver side of the circuit. The laser pen can be battery powered or wired to a power supply of equal voltage as the batteries.

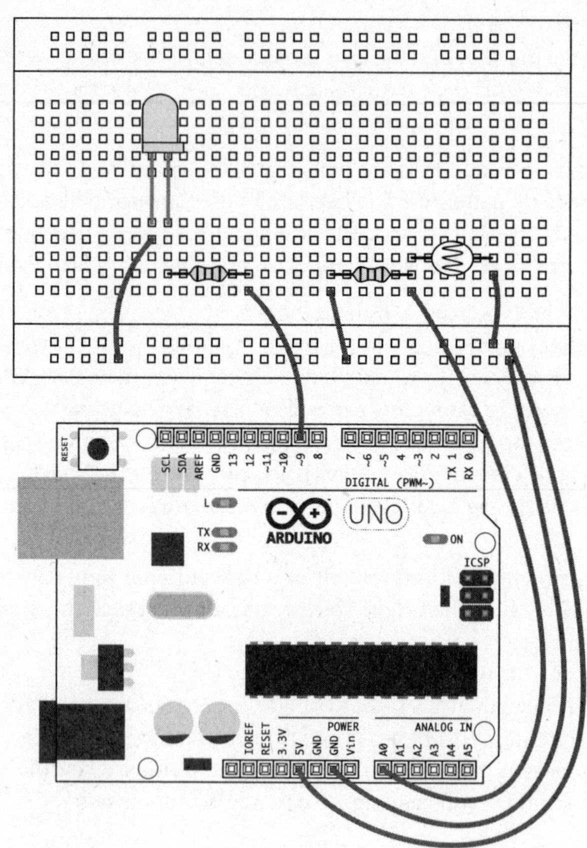

FIGURE 11-8: Analog input and LED output circuit layout.

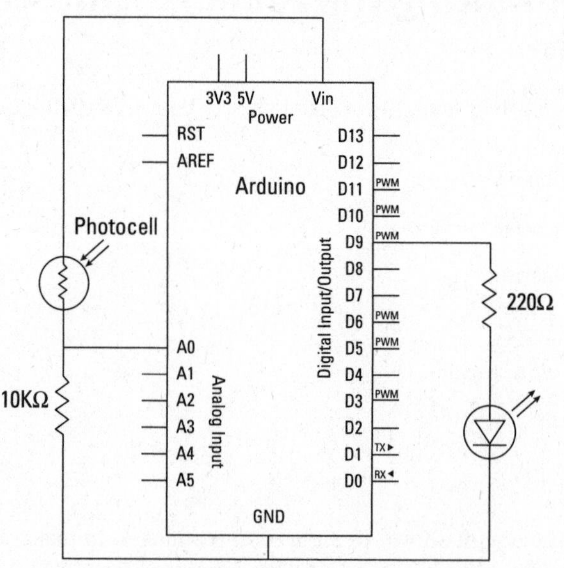

FIGURE 11-9: Analog input and LED output circuit diagram.

Choose File ➪ Examples ➪ 03.Analog ➪ AnalogInOutSerial from the Arduino menu to load the sketch.

```
/*
  Analog input, analog output, serial output

  Reads an analog input pin, maps the result to a range from 0 to 255 and
  uses the result to set the pulse width modulation (PWM) of an output pin.
  Also prints the results to the Serial Monitor.

  The circuit:
  - potentiometer connected to analog pin 0.
    Center pin of the potentiometer goes to the analog pin.
    side pins of the potentiometer go to +5V and ground
  - LED connected from digital pin 9 to ground

  created 29 Dec. 2008
  modified 9 Apr 2012
  by Tom Igoe

  This example code is in the public domain.

  http://www.arduino.cc/en/Tutorial/AnalogInOutSerial
*/

// These constants won't change.  They're used to give names
// to the pins used:
const int analogInPin = A0;   // Analog input pin that the potentiometer is
                              // attached to
const int analogOutPin = 9;   // Analog output pin that the LED is attached to

int sensorValue = 0;          // value read from the pot
int outputValue = 0;          // value output to the PWM (analog out)

void setup() {
  // initialize serial communications at 9600 bps:
  Serial.begin(9600);
}

void loop() {
  // read the analog in value:
  sensorValue = analogRead(analogInPin);
  // map it to the range of the analog out:
  outputValue = map(sensorValue, 0, 1023, 0, 255);
```

```
  // change the analog out value:
  analogWrite(analogOutPin, outputValue);

  // print the results to the serial monitor:
  Serial.print("sensor = " );
  Serial.print(sensorValue);
  Serial.print("\t output = ");
  Serial.println(outputValue);

  // wait 2 milliseconds before the next loop
  // for the analog-to-digital converter to settle
  // after the last reading:
  delay(2);
}
```

Press the Compile button to check your code. The compiler highlights any grammatical errors, which light up in red. If the sketch compiles correctly, click Upload to send the sketch to your board. When it has finished uploading, mount your laser so that it hits the center of the light sensor. Set the serial monitor to a baud rate of 9600 and you should see analog values at the high end of the range (1024 max). When you obstruct the beam, the range should drop and the LED should show the change. Experiment with the values in the `map` function to determine the best range of values.

When the value drops below a certain threshold, you can trigger a variety of actions — you have a highly sensitive trip wire sensor.

If nothing happens, double-check your wiring:

- Make sure that you're using the correct pin number.
- Check the connections on the breadboard. If the jump wires or components are not connected using the correct rows in the breadboard, they do not work.

Understanding the AnalogInOutSerial sketch

For more details on the workings of this sketch, see the notes in AnalogInOutSerial in Chapter 6. You can also find suggestions for different sketches to provide smoothing and calibration in Chapter 10.

Detecting Movement

A *passive infrared (PIR) sensor* is a common sensor in some homes and most commercial buildings. You may have seen this sensor in the corner of a room, blinking red every once in a while. It registers heat given off by people, animals, or other heat sources as infrared radiation. Infrared radiation is invisible to the human eye but is easy for the sensor to distinguish. The sensor itself is similar to the sensor found in a digital camera but without the complex lenses to capture a detailed picture. Essentially, a PIR sensor is somewhere between a high-resolution light sensor and a low-resolution camera. Simple lenses are usually fitted to PIR sensors to give them a wider viewing angle.

Most commonly, this type of sensor is used for motion detection in burglar alarms. Rather than detect motion, however, it detects changes in temperature. Temperature changes can trigger an alarm system or something more interesting (henchmen!), but the sensor is purely a way of monitoring changes in an environment.

You have two ways to get a PIR sensor. The first is to take apart a PIR burglar alarm, which is likely prepackaged with a lens and a sensor, which may be difficult to identify. The second method is to buy one specifically intended for microcontroller projects. This sensor most often comes with a basic, ping-pong-ball-styled lens and a bare circuit board underneath. The latter is easier to work with because all the details are known and it is described later in this section.

Consider the following during planning:

» **Complexity:** It can be tricky to hack an existing PIR sensor made for a specific system. Because it needs to communicate with that system, however, the sensor usually has clearly marked connections on the back. One of the benefits of using an existing sensor is that it is prepackaged, which reduces the amount of time you have to spend putting components together. Prepackaged systems are designed to be easy to install, so you may also be able to use manual calibration, by way of a potentiometer or a screwdriver slot, which can be a huge benefit for on-the-fly calibration rather than having to re-upload.

  If you're using a PIR sensor that is not prepackaged, it should be a lot more straightforward on the hardware and software side but requires careful thought concerning the housing. Some PIR sensors have their own on-board logic and operate like a switch, going `HIGH` when movement occurs over the threshold. This kind of sensor needs calibration to identify change from the norm.

- **Cost:** A household PIR sensor costs between $15 and $45. The main expense is the housing, which is usually designed to be discreet or look suitably high-tech. Bare PIR sensors cost a fraction of the price at around $10, but need a suitable housing to be of any real use.
- **Location:** Many housings allow you to neatly fit the sensor against a wall. Or you might consider using mini tripods for direction, as suggested in the previous section. Some of the tripod mounts also come with a suction-cup mount, which is perfect for fixing your sensor to smooth surfaces such as glass.

Most PIR sensors come ready for action, needing only power. They calibrate themselves based on what they can see and then send a `HIGH` or `LOW` value when they detect change. This makes them easy to program because you're dealing with the same signals as with a pushbutton.

Implementing the DigitalReadSerial sketch

In the DigitalReadSerial sketch example, you learn how to use the SE-10, a PIR sensor available from all major Arduino retailers. This PIR sensor has three wires: red, brown, and black. The red wire is the power source and should be connected to 5V. Oddly, the black wire is the signal wire and not the ground (see Figure 11-10; the black is the leftmost wire, brown is in the middle, and red is on the right). Brown should be wired to ground and black to pin 2.

FIGURE 11-10: The SE-10, with its strangely color-coded wires.

The signal pin is known as an open collector and needs to be pulled `HIGH` to start with. To do so, you use a 10k resistor to connect it to 5V as well. The pin, therefore, reads `HIGH` when no motion occurs and is pulled to `LOW` when there is motion.

You need the following:

- An Arduino Uno
- A breadboard
- An SE-10 PIR motion sensor
- A 10k ohm resistor
- Jump wires

Lay out the circuit as in the layout and circuit diagrams shown in Figures 11-11 and 11-12.

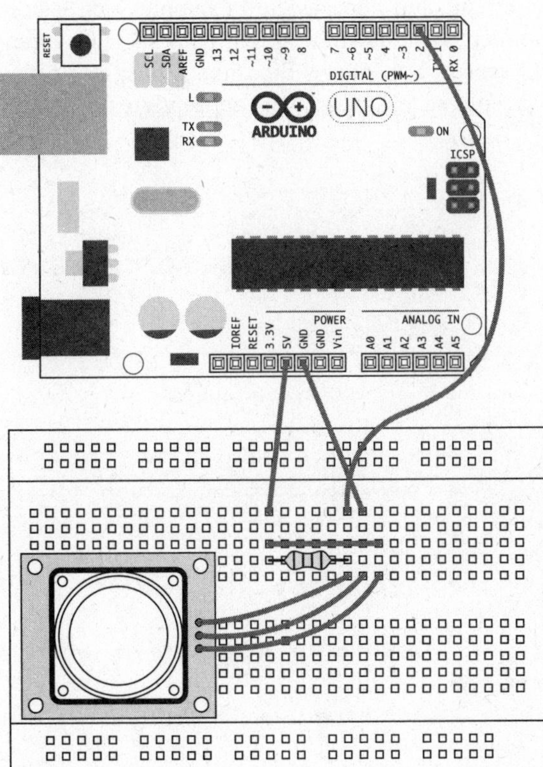

FIGURE 11-11: A PIR sensor circuit layout.

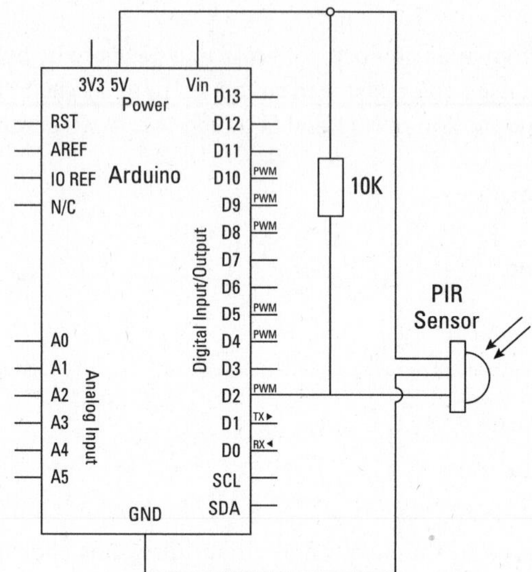

FIGURE 11-12: A PIR sensor circuit diagram.

Complete the circuit and choose File⇨ Examples⇨ 01.Basics⇨ DigitalReadSerial from the Arduino menu to load the sketch. This sketch is intended for a pushbutton, not a PIR sensor, but follows the same principles. If you want to make the sketch more specific, save it with a more appropriate file name and variable names.

```
/*
  DigitalReadSerial

  Reads a digital input on pin 2, prints the result to the Serial Monitor

  This example code is in the public domain.

  http://www.arduino.cc/en/Tutorial/DigitalReadSerial
*/

// digital pin 2 has a pushbutton attached to it. Give it a name:
int pushButton = 2;

// the setup routine runs once when you press reset:
void setup() {
  // initialize serial communication at 9600 bits per second:
  Serial.begin(9600);
  // make the pushbutton's pin an input:
  pinMode(pushButton, INPUT);
}

// the loop routine runs over and over again forever:
```

```
void loop() {
  // read the input pin:
  int buttonState = digitalRead(pushButton);
  // print out the state of the button:
  Serial.println(buttonState);
  delay(1);        // delay in between reads for stability
}
```

Press the Compile button to check your code. Doing so highlights any grammatical errors and turns them red. If the sketch compiles correctly, click Upload to send the sketch to your board. When it has finished uploading, affix the PIR sensor to a surface that is free of movement. Then open the serial monitor, which resets the sketch. The sensor calibrates itself in the first 1 to 2 seconds. When movement is detected, you should see the `buttonState` value change from 1 (no movement) to 0 (movement).

If nothing happens, double-check your wiring:

- » Make sure that you're using the correct pin number.
- » Check the connections on the breadboard. If the jump wires or components are not connected using the correct rows in the breadboard, they will not work.
- » Try restarting the PIR sensor by disconnecting and reconnecting the GND wire, and be sure that it does not move during or after calibration.

Understanding the DigitalReadSerial sketch

The only variable to declare is pin 2, the `pushButton` pin, or in this case the `pirSensor` pin:

```
// digital pin 2 has a pushbutton attached to it. Give it a name:
int pushButton = 2;
```

In `setup`, the serial port is opened and set to a baud rate of 9600; the input pin is set to an output:

```
// the setup routine runs once when you press reset:
void setup() {
  // initialize serial communication at 9600 bits per second:
  Serial.begin(9600);
  // make the pushbutton's pin an input:
  pinMode(pushButton, INPUT);
}
```

In loop (), the input pin is read and its value stored in buttonState. This value reads HIGH when no movement is occurring because the pull-up resistor is providing a voltage from the 5V pin. When there is movement, the open collector grounds the voltage and reads LOW:

```
// the loop routine runs over and over again forever:
void loop() {
  // read the input pin:
  int buttonState = digitalRead(pushButton);
```

The value of the input is then printed to the serial monitor:

```
  // print out the state of the button:
  Serial.println(buttonState);
  delay(1);        // delay in between reads for stability
}
```

TECHNICAL STUFF

This is one example of how to use existing code for different hardware. From this point, it is possible to trigger different outputs based on the HIGH or LOW signal of the PIR sensor. For ease of use and the clarity of other people who might use your code, rename variables to more appropriate names, add your own comments, and save the sketch so that you can easily distinguish it from the pushbutton sketch.

Measuring Distance

Two sensors for measuring distance that are extremely popular are the infrared proximity sensor and the ultrasonic range finder. They work in similar ways and achieve pretty much the same thing, but it's important to choose the right sensor for the environment you're in. An *infrared proximity sensor* has a light source and a sensor. The light source bounces infrared light off an object and back to the sensor, and the time it takes the light to return is measured to indicate the object's distance.

An *ultrasonic range finder* fires out high-frequency sound waves and listens for an echo when they hit a solid surface. By measuring the time it takes a signal to bounce back, the ultrasonic range finder can determine the distance traveled.

Infrared proximity sensors are not as accurate and have a much shorter range than ultrasonic range finders.

Consider the following during planning:

- **Complexity:** Both sensors are designed to be easy to integrate with Arduino projects. In the real world, they're used for similar electronics applications, such as proximity meters on the back of cars that beep as you approach the curb. Again, the main complexity is housing them effectively. Infrared proximity sensors such as those made by Shape have useful screw holes on the outside of the body of the sensor. MaxBotix makes ultrasonic range finders that don't have these mounts, but their cylindrical shape makes them simple to mount by drilling a hole through a surface or housing.
- **Cost:** Infrared proximity sensors cost in the region of $15 and have a range up to about 59 inches (150 cm). Ultrasonic range finders have a far greater possible range and accuracy but an equally great price, costing between $25 for a sensor that can read up to 254 inches (645 cm) and $100 for a more weather-resistant model that can read up to 301 inches (765 cm).
- **Location:** A common application for these sensors is monitoring the presence of a person or an object in a particular floor space, especially when a pressure pad would be too obvious or easy to avoid, or when a PIR sensor would measure too widely. Using a proximity sensor lets you know where someone is in a straight line from that sensor, making it a useful tool.

IR proximity sensors are okay in dark environments but perform terribly in direct sunlight. The MaxBotix ultrasonic range finder is one of my favorite and most reliable sensors. When using ultrasonic range finders, you can also choose how wide or narrow a beam you want. A large, teardrop-shaped sensor is perfect for detecting large objects moving in a general direction, whereas narrow beams are great for precision measurements.

Implementing the MaxSonar sketch

In this example, you learn how to measure precise distances using a MaxBotix LV-EZ0. The EZ0, EZ1, EZ2, EZ3, and EZ4 all work the same way, but each has a slightly narrower beam, so choose the appropriate one for your project.

The range finder needs some minor assembly. To use the range finder in your circuit, you need to solder on either header pins (to use the range finder on a breadboard) or lengths of wire.

You can connect your range finder by using analog, pulse width, or serial communication. In this example, you learn how to measure the pulse width and convert that to distance. The analog output can be read straight into your analog input pins, but it provides less accurate results than pulse width. This example does not cover serial communication.

You need the following for the MaxSonar sketch:

- » An Arduino Uno
- » An LV-EZ0 ultrasonic range finder
- » Jump wires

Complete the circuit from the layout and circuit diagrams in Figures 11-13 and 11-14. The connections for the range finder are clearly marked on the underside of the PCB. The 5V and GND connections provide power for the sensor and should be connected to the 5V and GND supplies on your Arduino. The PW connection is the pulse-width signal that will be read by pin 7 on your Arduino. Make sure that your distance sensor is affixed to some sort of base pointed in the direction that you want to measure.

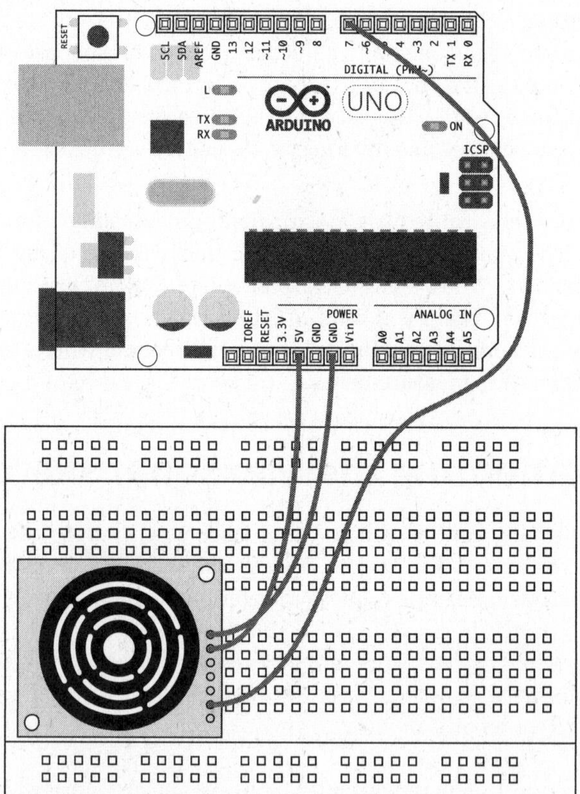

FIGURE 11-13: An LV-EZ0 circuit layout.

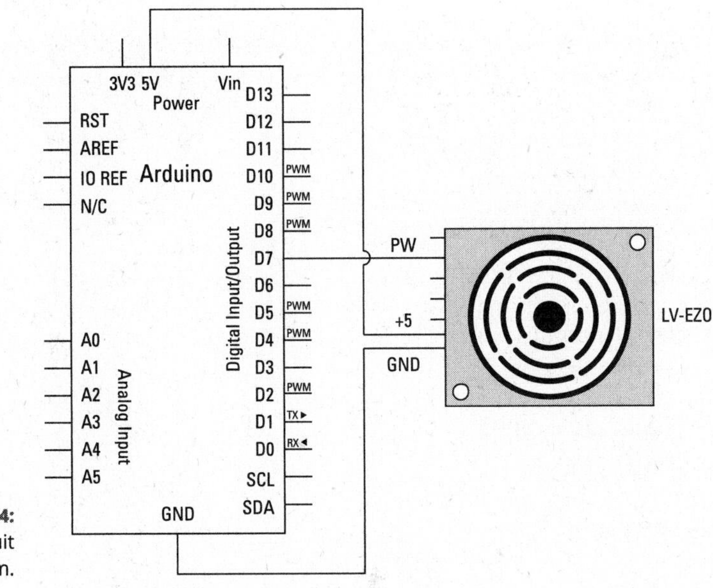

FIGURE 11-14: An LV-EZ0 circuit diagram.

You can find the MaxSonar code by Bruce Allen, along with some additional notes and functions, in the Arduino playground at `www.arduino.cc/playground/Main/MaxSonar`. Create a new sketch, copy or type the code into it, and save it with a memorable name, such as `myMaxSonar`:

```
//Feel free to use this code.
//Please be respectful by acknowledging the author in the code if you use or
  modify it.
//Author: Bruce Allen
//Date: 23/07/09
//Digital pin 7 for reading in the pulse width from the MaxSonar device.
//This variable is a constant because the pin will not change throughout
  execution of this code.
const int pwPin = 7;
//variables needed to store values
long pulse, inches, cm;

void setup() {
  //This opens up a serial connection to shoot the results back to the PC
  console
  Serial.begin(9600);
}

void loop() {

  pinMode(pwPin, INPUT);
```

```
  //Used to read in the pulse that is being sent by the MaxSonar device.
  //Pulse Width representation with a scale factor of 147 uS per Inch.

  pulse = pulseIn(pwPin, HIGH);
  //147uS per inch
  inches = pulse/147;
  //change inches to centimetres
  cm = inches * 2.54;

  Serial.print(inches);
  Serial.print("in, ");
  Serial.print(cm);
  Serial.print("cm");
  Serial.println();

  delay(500);

}
```

Press the Compile button to check your code. The compiler highlights any grammatical errors, turning them red. If the sketch compiles correctly, click Upload to send the sketch to your board. When it has finished uploading, open the serial monitor to see the distance measured in inches and centimeters. If the value is fluctuating, try using an object with a bigger surface.

This sketch allows you to accurately measure distance in a straight line. Check the results with a tape measure and make adjustments to the code if you find discrepancies.

If nothing happens, double-check your wiring:

- » Make sure that you're using the correct pin number.
- » Check the connections on the breadboard. If the jump wires or components are not connected using the correct rows in the breadboard, they will not work.

Understanding the MaxSonar sketch

In the declarations, pin 7 is defined as the `pwPin`:

```
//This variable is a constant because the pin will not change throughout
   execution of this code.
const int pwPin = 7;
```

`long` variables are used to store the pulse width and distances in inches and centimeters. Note that you can declare all three in a list if they have no value:

```
//variables needed to store values
long pulse, inches, cm;
```

In `setup`, the serial connection is opened to print the results:

```
void setup() {
  //This opens up a serial connection to shoot the results back to the PC
   console
  Serial.begin(9600);
}
```

In the main `loop ()`,`pwPin` is set as an input. You can set the input in `loop ()` or move it to `setup ()`:

```
void loop() {

  pinMode(pwPin, INPUT);
```

You use the `pulseIn` function to return the length of time it takes the pulse to return to the sensor in microseconds, or µS:

```
    //Used to read in the pulse that is being sent by the MaxSonar device.
  //Pulse Width representation with a scale factor of 147 uS per Inch.

  pulse = pulseIn(pwPin, HIGH);
```

A pulse travels 1 inch every 147 µS, so you can calculate the number of inches based on the time. From this information, a simple conversion outputs the distance in different units:

```
  //147uS per inch
  inches = pulse/147;
  //change inches to centimetres
  cm = inches * 2.54;
```

The results are printed to the serial monitor, with `Serial.println` at the end to start a new line in between each reading:

```
  Serial.print(inches);
  Serial.print("in, ");
  Serial.print(cm);
```

```
Serial.print("cm");
Serial.println();
```

A delay is added to slow the readings for legibility, but you can remove the delay if responsiveness is more important:

```
delay(500);

}
```

The preceding code provides you with an accurate distance reading that you can incorporate into your own projects. A simple way to make use of this reading is with an `if` statement. For example:

```
if (cm < 50) {
// do something!
}
```

Testing, Testing . . . Can Anybody Hear This?

Sound is another way to detect presence, and the best way to do it is with an electret microphone. It's common to think of sounds in their analog form as recognizable noises, but a lot of the sounds you hear every day have undergone an analog-to-digital conversion. By converting sound into a digital signal, it's possible to interpret it on a computer or an Arduino. An *electret microphone* is similar to the microphones found in computer headsets and is extremely sensitive, but it needs an amplifier for the Arduino to register the readings.

Consider the following during your planning:

- **Complexity:** You can choose from plenty of electret mics, but by far the easiest is SparkFun's electret microphone breakout. It comes preassembled with the mic mounted on a circuit board and with an amplifier, and can be easily wired to an Arduino as an analog input. It's possible to use electret mics from other headsets or desktop microphones, but these need their own amplifier to be of any use. Some work is involved to make the correct housing for the mic to protect it from the environment or human contact, but this housing could be a simple enclosure with a hole.

- **Cost:** The microphone itself is extremely cheap at 90 cents from SparkFun (or distributors of their products). The breakout board costs $5.95, which is not a huge expense for the amount of labor saved.
- **Location:** As an ambient sensor, the mic could be placed just about anywhere to map the noise levels in a room. If you're listening for a specific noise such as a door slamming, it may be best to place the microphone near that source to get a clear reading.

One unusual use for a mic is monitoring someone's breath. Because the mic measures the amplitude or volume of the sound, it's perfect for this application. By placing the mic at the end of a tube, it's even possible to monitor the length and intensity of breaths as the air rushes past the mic.

Electret mics are great for measuring the amplitude or volume of noise, which can be used to trigger a variety of outputs.

Implementing the AnalogInOutSerial sketch

In this example, you monitor sound levels wherever you are using an electret mic. This simple sensor can be read as an analog input into your Arduino.

You need the following:

- An Arduino Uno
- An electret microphone breakout board
- Jump wires

Complete the circuit from the layout and circuit diagrams in Figures 11-15 and 11-16 to connect the mic as your input. The electret microphone breakout board requires a small amount of soldering to use it on a breadboard or connect it to your Arduino. You can solder on either a set of three header pins or a length of wire, depending on your situation.

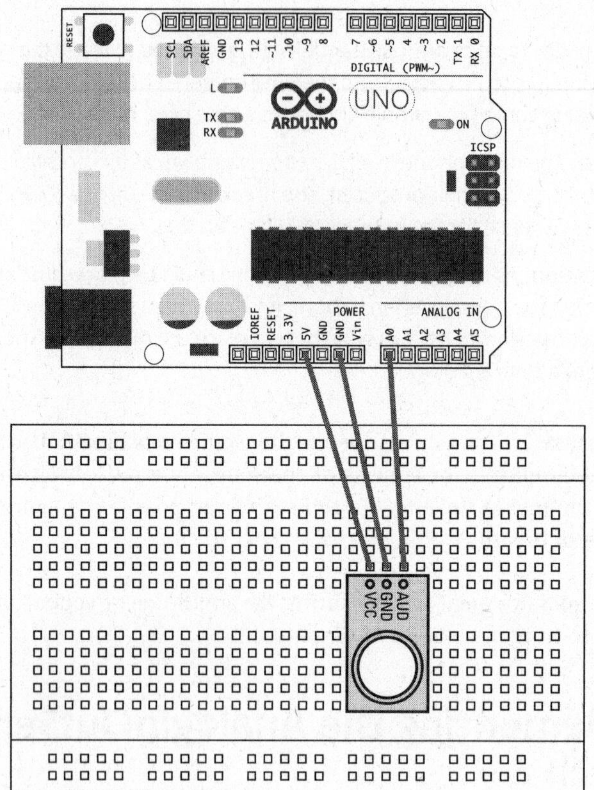

FIGURE 11-15: An electret microphone circuit layout.

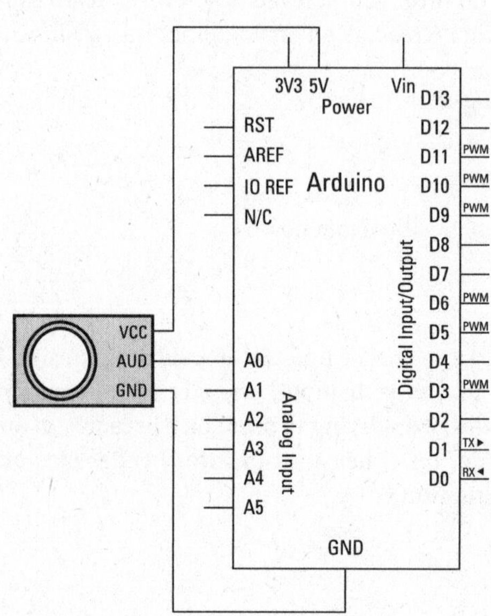

FIGURE 11-16: An electret microphone circuit diagram

Choose File ➪ Examples ➪ 03.Analog ➪ AnalogInOutSerial from the Arduino menu to load the sketch.

```
/*
  Analog input, analog output, serial output

  Reads an analog input pin, maps the result to a range from 0 to 255 and
  uses the result to set the pulse width modulation (PWM) of an output pin.
  Also prints the results to the Serial Monitor.

  The circuit:
  - potentiometer connected to analog pin 0.
    Center pin of the potentiometer goes to the analog pin.
    side pins of the potentiometer go to +5V and ground
  - LED connected from digital pin 9 to ground

  created 29 Dec. 2008
  modified 9 Apr 2012
  by Tom Igoe

  This example code is in the public domain.

  http://www.arduino.cc/en/Tutorial/AnalogInOutSerial
*/

// These constants won't change.  They're used to give names
// to the pins used:
const int analogInPin = A0;  // Analog input pin that the potentiometer is
   attached to
const int analogOutPin = 9; // Analog output pin that the LED is attached to

int sensorValue = 0;        // value read from the pot
int outputValue = 0;        // value output to the PWM (analog out)

void setup() {
  // initialize serial communications at 9600 bps:
  Serial.begin(9600);
}

void loop() {
  // read the analog in value:
  sensorValue = analogRead(analogInPin);
  // map it to the range of the analog out:
  outputValue = map(sensorValue, 0, 1023, 0, 255);
```

```
  // change the analog out value:
  analogWrite(analogOutPin, outputValue);

  // print the results to the serial monitor:
  Serial.print("sensor = " );
  Serial.print(sensorValue);
  Serial.print("\t output = ");
  Serial.println(outputValue);

  // wait 2 milliseconds before the next loop
  // for the analog-to-digital converter to settle
  // after the last reading:
  delay(2);
}
```

Press the Compile button to check your code. Doing so highlights any grammatical errors and turns them red. If the sketch compiles correctly, click Upload to send the sketch to your board. When it has finished uploading, open the serial monitor to see analog values in the range of 0 to 1024.

If nothing happens, double-check your wiring:

- » Make sure that you're using the correct pin number.
- » Check the connections on the breadboard. If the jump wires or components are not connected using the correct rows in the breadboard, they will not work.

See what range of values you get from different noises in your environment and how sensitive or overly sensitive the mic is. Another sketch to consider is the Smoothing sketch in Chapter 10.

Understanding the AnalogInOutSerial sketch

For more details on the workings of this sketch, see the notes in AnalogInOutSerial in Chapter 6. You can also find suggestions for different sketches to provide smoothing and calibration in Chapter 10.

» Looking at the range of shields available

» Understanding libraries

Chapter 12
Becoming a Specialist with Shields and Libraries

The further you progress in learning about Arduino, the more you want to do, and it's natural to want to run before you can walk. The areas that interest you may be highly specialized in themselves and require a huge investment of time to understand. Perhaps the most important thing about Arduino is the Arduino community, which is where you can get help when you want go further.

The traditional viewpoint that is hammered into us in education is to protect our ideas for dear life. Thankfully, many people in the Arduino community have seen past that limitation and are kind enough to share their hard work. By sharing this knowledge, the Arduino community helps the hardware and software become available to more people, who find new and interesting uses for it. If these people in turn share their results, the community continues to grow and eventually makes even the most difficult projects achievable. In this chapter, you discover the power of shared resources such as shields and libraries, even for beginners.

Looking at Shields

Shields are pieces of hardware that sit on top of your Arduino, often to give it a specific purpose. For example, you can use a shield to make it easier to connect and control motors or even to turn your Arduino into something as complex as a mobile phone. A shield may start out as an interesting bit of hardware that an enthusiast has been experimenting with and wants to share with the community. Or an enterprising individual (or company) may design a shield to make an application easier based on demand from the Arduino community.

Shields can be simple or complex. They are sold preassembled or as kits. Kits allow you more freedom to assemble the shield as you need it to be. Some kits require you to assemble the circuitry of the boards, although more complex shields may already be largely assembled, needing only header pins.

Shields enable you to use your Arduino for more than one purpose and to change that purpose easily. They neatly package the electronics for that circuit in the same footprint as an Arduino, and are stackable to combine different functionalities. But they all have to use the same pins on the Arduino, so if you stack shields, watch out for those that need to use the same pins. They always connect the GND pins, too, because any communication by your Arduino and another device needs a common GND.

Considering combinations

In theory, shields could be stacked on top of each other forever, but you should take some points into consideration before combining them:

- **Physical size:** Some shields just don't fit on top of one another. Components that are higher than the header sockets may touch the underside of any board on top of it. This situation, which can cause short circuits if a connection is made that shouldn't be, can seriously damage your boards.
- **Obstruction of inputs and outputs:** If an input or output is obstructed by another shield, it becomes redundant. For example, there's no point having a joystick shield or an LCD shield under another shield because no more than one can be used.
- **Power requirements:** Some hardware requires a lot of power. Although it is all right for shields to use the same power and ground pins, there is a limit to the amount of current that can flow through the other input/output (I/O) pins: 40mA per pin and 200mA max between all I/O pins. Exceed this, and you run the risk of seriously damaging your board and any other attached shield. In most cases, you can easily remedy this problem by powering your Arduino

and shields from an external power supply so that the current isn't passed through the Arduino. Make sure to use a common GND if you're communicating between a board using I2C, SPI, or serial.

- **Pins:** Some shields require the use of certain pins. It's important to make sure that shields aren't doubling up on the same pins. In the best case, the hardware will just be confused; in the worst case, you can send voltage to the wrong place and damage your board.
- **Software:** Some of these shields need specific libraries to work. There can be conflicts in libraries calling on the same functions, so make sure to read up on what's required for your shield.
- **Interference with radio/Wi-Fi/GPS/GSM:** Wireless devices need space to work. Move antennas or aerials away from the board to get a clear signal. If an antenna is mounted on the board, it's generally a bad idea to cover it. Always try to place wireless shields at the top of the stack.

Reviewing the field

To give you an idea of available shields, this section covers some of the most interesting and useful shields on the market and shows you where to look for more information.

TIP

Note that all prices were current at the time this book was written and are liable to change, but I've included them to give you an idea of the cost. Links to products may also change, so always try searching for the product if the link is broken. Technical information is gathered from the manufacturers' websites, but you should always check the details yourself to make sure that you're buying what you need. Boards are revised occasionally, so always keep an eye out for the latest versions.

Finally, a lot of feedback on all of these products is available online, so always read the comments and forums to get a good understanding of what you're buying.

This range of shields covers a vast number of different uses and the huge potential of Arduino projects. For many projects, a shield is all you need, but a shield is also an excellent stepping-stone for proving a concept before refining or miniaturizing your project.

REMEMBER

Prices provided are from a range of distributors to show the approximate value of the items. If you are a savvy shopper or are looking to buy in bulk, you may be able to reduce the cost.

Proto shield kit Rev3

Made by: Arduino

Price: $9.90 from Arduino Store

Pins used: None

The Proto shield (shown in Figure 12-1) is a platform for building custom circuits on your Arduino. Many shields listed in this chapter add a specific function to your Arduino, but with a Proto Shield, you can decide how to use it. Take your existing breadboard layouts and solder them to the surface of the Proto shield to make your project more durable. Proto shields also come in a larger size to match the Arduino Mega's footprint. Another handy feature of these shields is the space to attach SMD parts, which can be difficult to do otherwise.

FIGURE 12-1: A fully assembled Proto shield.

The Proto shield is sold either fully assembled or as a kit that requires soldering.

You can find details about the shield on the Arduino product page (`store.arduino.cc/usa/proto-shield-rev3-uno-size`).

ProtoScrew shield

Made by: WingShield Industries

Price: $14.95 from Adafruit

Pins used: None

ProtoScrew shield is similar to the regular Proto shield but has large screw terminals connected to the pins as well. This feature is great for applications that have lots of inputs that may need changing or swapping, or just for easier assembly and disassembly. Changing a piece of wire is much easier with screw terminals than with soldering, so bear this in mind when planning your next project.

ProtoScrew shield is sold as a kit and requires soldering. You can find more details on the SparkFun products page (`adafruit.com/product/196`).

Adafruit Wave shield v1.1

Made by: Adafruit

Price: $22.00 from Adafruit

Pins used: 13, 12, 11, 10, 5, 4, 3, 2 on the Uno R3

The Wave shield (see Figure 12-2) is a relatively cheap kit that allows you to play sounds or music with your Arduino. The Wave shield allows you to play WAV files directly from an SD card, making it easy to upload and change the sound files from your computer. To use the shield, you need the WaveHC library, which is available from the product page and Google Code (`http://code.google.com/p/wavehc/`).

The Wave shield is sold as a kit and requires soldering. The SD card reader must use pins 13, 12, and 11 because they support a high-speed Serial Peripheral Interface (SPI), which is a protocol needed to transfer data quickly. Pin 10 is used to communicate with the SD card reader, and pins 5, 4, 3, and 2 are used to talk to the digital-to-analog converter (DAC), which converts a digital music signal into an analog voltage.

For more details, visit the product page on Adafruit's website (`www.adafruit.com/products/94`).

FIGURE 12-2: A fully assembled Wave shield.

MP3 player shield

Made by: SparkFun

Price: $24.95 from SparkFun

Pins used: 13, 12, 11, 9, 8, 7, 6, 5, 4, 3, 2 on the Uno R3

Turn your Arduino into an MP3 player with the easy-to-assemble MP3 player shield from SparkFun! Not only can it decode MP3 files, it's also capable of decoding Ogg Vorbis, AAC, WMA, and MIDI. The MP3 shield (shown in Figure 12-3) also has a microSD card reader for ease of uploading files, and it has a 3.5mm mini jack that you can connect to most speaker systems.

The MP3 player shield is assembled but requires minor soldering to attach the header pins or header sockets. The SD card reader uses pins 13, 12, and 11. You use pin 9 to talk with the SD card reader. Use pins 8, 7, 6, and 2 talk to MP3 audio decoder VS1053B, and use pins 4 and 3 for additional MIDI functionality.

For more details, visit the SparkFun products page at `www.sparkfun.com/products/10628`. Also see the tutorial page at `learn.sparkfun.com/tutorials/mp3-player-shield-hookup-guide-v15`. The comments below the tutorial address many of the issues with this guide. One user has even written a library to make your life easier — a great example of the Arduino community supporting existing products.

FIGURE 12-3: An MP3 shield kit.

REMEMBER

Always read the comments and forum entries on products and kits. These comments often contain a lot of detail on the ease of (or difficulty with) a product. This is also the place to voice your own problems. Just be sure that you're not repeating something that's solved further down the page; otherwise, you'll be advised to read the manual!

MIDI shield

Made by: SparkFun

Price: $19.95 from SparkFun

Pins used: Uses pins 2, 3, 4, 6, 7, A1, A0 on the Uno R3

MIDI (Music Instrument Digital Interface) revolutionized the music industry in the 1980s and is still widely used now to connect instruments, computer, stage effects, and other hardware. With the MIDI shield, you can interface with anything that can send or receive MIDI data and incorporate it into your Arduino project.

The MIDI shield is sold as a kit and requires soldering. For more details, visit the SparkFun product page (`www.sparkfun.com/products/12898`). You can find some excellent general tutorials on MIDI at `arduino.cc/en/Tutorial/Midi` and `itp.nyu.edu/physcomp/Labs/MIDIOutput` and a lot of excellent reference material at `www.tigoe.net/pcomp/code/communication/midi/` and `hinton-instruments.co.uk/reference/midi/protocol/`.

RGB LCD shield with 16 x 2 character display

Made by: Adafruit

Price: $24.95 from Adafruit

Pins used: Uses pins A4 and A5 on the Uno R3

This handy LCD (liquid crystal display) shield packages everything you need onto one board. LCDs are found in older mobile phones and Nintendo GameBoys (wow, that sounds old). They use a film that sits over a solid-colored, usually backlit surface. The pixels of this film can be turned on or off to make shapes, text, or graphics, and this is what you control with your Arduino. At the center of the shield is an RGB LCD display, so instead of being stuck with just one color, you can choose from *any* RGB color.

The RGB backlight is controlled directly from your Arduino. The display is a 16 x 2 character display (no graphics), which means you can write two rows of 16 characters. You can choose from two displays: colored text on a dark background (negative) or dark text on a colored background (positive). A variety of LCD character displays with various backlighting and dimensions is available, so be sure to shop around.

The RGB LCD shield is sold as a kit and requires soldering. Instead of using nine pins or more, the LCD, backlight, and buttons together use just two. By using the I2C to communicate with the shield, you can use only analog pin 4, which is the data (SDA) line, and analog pin 5, which is the clock (SCL) line. This protocol is used in many devices, so it is extremely useful to know about it. For more details on I2C, check John Boxall's excellent tutorial at `tronixstuff.wordpress.com/2010/10/20/tutorial-arduino-and-the-i2c-bus/`.

For more details, check out the Adafruit product page at `adafruit.com/products/714` and tutorial at `learn.adafruit.com/rgb-lcd-shield`.

Shields are also available that use the same technology but don't limit you to letters and numbers. If you're looking to display your own graphics, you might want to use the SparkFun color LCD shield, which uses a Nokia 6100 screen, or the larger TFT touch shield.

2.8" TFT touch shield with capacitive touch

Made by: Adafruit

Price: $44.95 from Adafruit

Pins used: 4, 8, 9, 10, 11, 12, 13, A4, A5

If an LCD display isn't enough for you, try the TFT touch shield to add full color and touch input to your project. This display is a TFT LCD screen — a variation on a standard LCD screen that uses thin-film transistor (TFT) technology to improve the image quality — with a resolution of 240 x 320 pixels and 18-bit colors, giving you 262,144 shades. The screen is also fitted with a resistive touchscreen to register finger presses anywhere on the surface of the screen.

The TFT touch shield is sold fully assembled and requires no soldering, so you can simply plug it on top of your Arduino. The touch shield needs a lot of pins to function and leaves you with only digital pins 2 and 3 and analog pins 4 and 5. Pin 12 is also available if you're not using the microSD reader.

Check out the products page at `www.adafruit.com/products/376` and the full tutorial at `learn.adafruit.com/adafruit-2-8-tft-touch-shield-v2`. Adafruit has also kindly written a complete library for the TFT to draw pixels, shapes, and text (`github.com/adafruit/TFTLCD-Library`) and a library for the touchscreen that detects x, y, and z horizontal movement, vertical movement, and pressure (`github.com/adafruit/Adafruit_TouchScreen`).

Joystick shield

Made by: SparkFun

Price: $12.95 from SparkFun

Pins used: 2, 3, 4, 5, 6, A0, A1

The joystick shield (shown in Figure 12-4) has all the functions of a modern game controller on a single Arduino-compatible board. It provides not only four pushbuttons to assign to various functions but also a hidden button in the control stick itself. With the ergonomic control stick, you can smoothly transition between x and y axes to perform movements with great accuracy.

The joystick shield is sold as a kit and requires soldering. It uses only five digital pins and two analog pins, leaving many other Arduino pins free for other uses. It has five pushbuttons, using digital pins 2 to 6. The movement of the joystick is measured using two potentiometers: analog 0 maps the *x*, or horizontal, movement; analog 1 maps the *y*, or vertical, movement.

You can find more details on the SparkFun product page (`www.sparkfun.com/products/9760`). You can also see in-depth assembly tutorial (`www.sparkfun.com/tutorials/161`) and quick-start guide (`www.sparkfun.com/tutorials/171`).

FIGURE 12-4: A joystick shield.

Adafruit motor/stepper/servo shield kit v2.3

Made by: Adafruit

Price: $19.95 from Adafruit

Pins used: 9, 10, A4, A5

Love motors? Want to try them all? Then the Adafruit motor/stepper/servo shield is the one for you. The shield is aptly named, enabling you to run all those motors you love. You can connect up to two 5V hobby servos, two stepper motors, or four bidirectional DC motors. The screw terminals make attaching and switching motors easy. When dealing with motors, it's always important to make sure that you have enough current to drive them all, so a handy screw terminal on the shield allows you to power your motors independently of your Arduino.

The Adafruit motor/stepper/servo shield is sold as a kit and requires soldering. Pins A4 and A5 control any connected DC or stepper motors, and pins 9 and 10 control any connected servos. This leaves you with a lot of pins free to be used for other inputs or outputs.

You can find many details on the Adafruit product page (`www.adafruit.com/product/1438`) and in the in-depth tutorial (`learn.adafruit.com/adafruit-motor-shield-v2-for-arduino/`). Be aware of how much load is on the motor because the shield is designed to provide up to 600 mA per motor, with 1.2A peak current. If you're approaching 1A, include a heat sink on the motor driver to dissipate the heat.

Also, the nice people at Adafruit provide an easy-to-use library for your motor project (github.com/adafruit/Adafruit_Motor_Shield_V2_Library). Happy motoring!

LiPower shield

Made by: SparkFun

Price: $29.95 from SparkFun

Pins used: 3

If you want to make your Arduino project more mobile, batteries are the answer. Rather than use bulky AA or AAA battery packs, the LiPower shield allows you to use rechargeable lithium batteries instead. Although lithium batteries are rated as only 3.7V, some clever hardware steps them up to 5V to make them sufficient to power your Arduino.

The LiPower shield is assembled but requires minor soldering to attach the header pins or header sockets. Because the LiPower shield is there to provide power and not consume it, only one pin is in use. Pin 3 can be configured as an alert interrupt pin to signal whenever the battery drops to 32 percent or lower.

For more details, check out the SparkFun product page (www.sparkfun.com/products/13158). You find interesting notes on the hardware regarding the difficulties with charging lithium batteries, so make sure to read all the comments below the product description.

Many other smaller lithium breakout boards are available that supply the standard 3.7V, such as the SparkFun USB LiPoly charger (www.sparkfun.com/products/12711) and the Adafruit micro Lipo (www.adafruit.com/product/1904). These breakout boards are perfect when paired with an appropriate low-voltage Arduino, such as the Arduino MKR1000 (store.arduino.cc/usa/arduino-mkr1000). Boards such as the Arduino MRK1000 are useful when trying to reduce the size of your Arduino project.

Adafruit Ultimate GPS logger shield

Made by: Adafruit

Price: $44.95 from Adafruit

Pins used: 7, 8, 10, 11, 12, 13, A4, A5

The Adafruit Ultimate GPS logger shield lets you track and store location information using the Global Positioning System. You can find your location within a few meters. Use it to create GPS art or to map all your movements over the month. It's also great for giving you an extremely accurate time. Data is stored on a microSD card as a TXT file, which can then be overlaid onto Google Maps or visualized in some other way.

With the ever-increasing storage capacity of SD cards, you can store much more data on an SD card than your Arduino can on its own internal memory. This extra capacity is especially useful because it keeps your data-logging device mobile without the need for a computer, so you can leave that bulky laptop at home and send your GPS device out into the world!

The Adafruit Ultimate GPS logger shield is sold without attached header pins and requires soldering. Pins 10, 11, 12, and 13 are used for communication with the SD card. The GPS module uses either the hardware serial connection (pins 0 and 1) or two pins for a software serial connection (pins 7 and 8). You can enable other optional functions, such as a signaling LED to indicate when data is logged, a pin to monitor the pulse from the GPS synchronization clock, and a pin to detect when an SD card is in the slot.

Get the shield and more information on Adafruit (`www.adafruit.com/product/1272`). You can view an in-depth tutorial on Adafruit's site (`learn.adafruit.com/adafruit-ultimate-gps-logger-shield/`), which details everything from construction of the kit to the Arduino code, along with uses for the GPS data.

Adafruit FONA 800 shield

Made by: Adafruit

Price: $39.95 from Adafruit

Pin requirements: 2, 3, 4

The Adafruit FONA 800 shield turns your modest Arduino into a functional mobile phone. With this shield, you can send and receive calls, text messages, and even data. All you need is a prepaid SIM card and an antenna, and you're ready to communicate with the world. By using serial commands you can send codes to talk to the SIM800L module.

The Adafruit FONA 800 shield is sold without attached header pins. You also need to purchase an antenna with an SMA connector; Adafruit offers a Slim Sticker-type GSM/Cellular quad-band antenna (`www.adafruit.com/product/3237`). The Arduino talks with the SIM800L via pins 2 and 3 using the SoftwareSerial library.

The SIM card should have enough credit to perform the actions you're trying to do. Those offering unlimited text messages are especially useful. Other optional extras are a mic and a speaker; without them, you won't be able to do anything more than call and hang up.

You can view an in-depth tutorial on Adafruit's site (`learn.adafruit.com/adafruit-fona-800-shield`).

Geiger counter – radiation sensor board shield

Made by: Liberium

Price: $170 from Cooking Hacks

Pin requirements: 2, 3, 4, 5, 6, 7, 8, 9, 10, 11, 12, 13

The radiation sensor board is probably one of the most impressive Arduino shields. It allows you to monitor radiation levels in the environment. This board was made to help the people of Japan monitor radiation levels following the radiation leakages in Fukushima in March 2011. The Geiger counter can use various Geiger tubes to detect different types and levels of radiation. There is also an LCD display, an LED, and a piezo speaker for feedback.

WARNING

This shield uses Geiger tubes that operate at dangerously high voltages (400V–1000V), so it requires extreme care. It is best to keep the radiation sensor board in an enclosure to keep it out of human contact. Radiation is dangerous, but so is electricity. If you don't know what you're doing, don't mess around.

The piezo speaker and LED are connected to pin 2, which triggers an interrupt with every pulse that the Geiger tube generates. Depending on the tube used and the number of pulses or counts per minute (cpm), you can determine the radiation level in Sieverts per hour. Pins 3 to 8 are used for the LCD display to generate the sensor readings in detail. Pins 9 to 13 are used for the LED bar to give clear visual feedback of the radiation level. The first three LEDs are green. The last two are red, showing that a high and potentially dangerous level of radiation is being approached.

More details on this project can be found on the Cooking Hacks product page at `www.cooking-hacks.com/index.php/documentation/tutorials/geiger-counter-arduino-radiation-sensor-board`.

Staying current

Many other great shields are available, and new and improved ones are released all the time. You can take some actions to keep up to date, though.

Check the stores regularly for the latest products. It's a bit like browsing in a DIY shop; you never know what you'll find:

- **Arduino Store** (`store.arduino.cc`)
- **Adafruit** (`www.adafruit.com`)
- **Maker Shed** (`www.makershed.com`)
- **Seeed Studio** (`www.seeedstudio.com`)
- **SparkFun** (`www.sparkfun.com`)

Also, check Arduino-related blogs regularly. Blogs and news pages on Arduino-related sites often show off new kits or new and interesting applications of older hardware, so they're always worth checking for inspiration:

- **Arduino Blog** (`arduino.cc/blog`)
- **Adafruit** (`www.adafruit.com/blog`)
- **Hack A Day** (`hackaday.com`)
- **Make** (`blog.makezine.com`)
- **Seeed Studio** (`www.seeedstudio.com/blog`)
- **SparkFun** (`www.sparkfun.com/news`)

Some people have made efforts to document the various shields. Check the Arduino shield list at `http://shieldlist.org`.

Browsing the Libraries

Basic sketches can get you quite a long way, but when you become more advanced you need to know about libraries. *Libraries* provide extra functionality to your sketch, either to use specific hardware or to incorporate more complex functions in software. In the same way that you'd go to a physical library to learn something new, you include libraries in your code to teach your Arduino something new. By including a library in a sketch, you can quickly and easily access functions to help you achieve your goals.

Getting started with complex hardware or software can be difficult. Luckily, a lot of people have taken the time to document their progress and have released libraries, often with examples, that you can easily integrate into your own sketches. From this, it's possible to get something working and gain a better understanding of it. This learn-by-doing approach of Arduino allows you to make a lot of progress quickly and easily with hardware or software that would otherwise be a huge challenge.

Reviewing the standard libraries

This section covers a selection of the libraries included in the current release of Arduino at the time of writing (1.8.5). The standard libraries cover a wide range of subject areas and are usually popular topics that have been heavily documented. You can find these libraries by choosing Sketch⇨Include Library⇨Manage Libraries.

Choosing a library includes one line at the top of your current sketch, such as `#include <EEPROM.h>`. Before you attempt to understand a library, you should try an example of it. You'll find examples at the bottom of the menu that appears at when you choose File⇨Examples.

Here is a brief description of what each library does:

- **EEPROM** (`arduino.cc/en/Reference/EEPROM`): Your Arduino has electronically erasable programmable read-only memory (EEPROM), which is permanent storage similar to the hard drive in a computer. Data stored in this location stays there even if your Arduino is powered down. Using the EEPROM library, you can read from and write to this memory.
- **Ethernet** (`arduino.cc/en/Reference/Ethernet`): After you have your Ethernet shield, the Ethernet library allows you to quickly and easily start talking to the Internet. When you use this library, your Arduino can act either as a server that is accessible to other devices or as a client that requests data.
- **Firmata** (`arduino.cc/en/Reference/Firmata`): Firmata is one way to control your Arduino from software on a computer. It is a standard communication protocol, so you can use the library to allow easy communication between hardware and software rather than write your own communication software.
- **LiquidCrystal** (`arduino.cc/en/Reference/LiquidCrystal`): The LiquidCrystal library helps your Arduino talk to most liquid crystal displays (LCDs). The library is based on the Hitachi HD44780 driver, and you can usually identify these displays by their 16-pin interface.
- **SD** (`arduino.cc/en/Reference/SD`): The SD library allows you to read from and write to SD and microSD cards connected to your Arduino. SD cards need to use SPI to transfer data quickly, which happens on pins 11, 12, and 13. You also need to have another pin to select the SD card when it's needed.

- **Servo** (`arduino.cc/en/Reference/Servo`)**:** The Servo library allows you to control up to 12 servo motors on the Uno R3 (and up to 48 on the Mega). Most hobby servos turn 180 degrees, and using this library, you can specify the degree that you want your servo(s) to turn to.
- **SPI** (`arduino.cc/en/Reference/SPI`)**:** The Serial Peripheral Interface (SPI) is a method of communication that allows your Arduino to communicate very quickly with one or more devices over a short distance. Example of this communication include receiving data from sensors, talking to peripherals such as an SD card reader, and communicating with another microcontroller.
- **SoftwareSerial** (`arduino.cc/en/Reference/SoftwareSerial`)**:** The SoftwareSerial library allows you to use any digital pins to send and receive serial messages instead of, or in addition to, the usual hardware pins (0 and 1). This capability is great if you want to keep the hardware pins free for communication to a computer, allowing you to have a permanent debug connection to your project while still being able to upload new sketches or to send duplicate data to multiple serial devices.
- **Stepper** (`arduino.cc/en/Reference/Stepper`)**:** The Stepper library allows you to control stepper motors from your Arduino. This code also requires the appropriate hardware to work, so make sure to read Tom Igoe's notes on the subject at `www.tigoe.net/pcomp/code/circuits/motors/stepper-motors/`.
- **Wi-Fi** (`arduino.cc/en/Reference/WiFi`)**:** The WiFi library is based on the Ethernet library listed previously, but with alterations specific to the Wi-Fi shield to allow you to wirelessly connect to the Internet. The WiFi library also works well with the SD library, allowing you to store data on the shield.
- **Wire** (`arduino.cc/en/Reference/Wire`)**:** The Wire library allows your Arduino to communicate with I2C devices (also known as TWI, or two-wire interface). Such devices could be addressable LEDs or a Wii Nunchuk, for example.

Installing additional libraries

Many libraries aren't included in the Arduino software by default. Some libraries are for unique applications such as specific hardware or functions; others are refinements or adaptations of existing libraries. Luckily, Arduino makes including these easy, so you can quickly try them all to see which are right for your needs.

Most libraries are easily installed by choosing Sketch ⇨ Include Library ⇨ Manage Libraries from the Arduino IDE menu. The Library Manager dialog box displays an easily searchable list of libraries submitted to the Arduino servers and approved for release.

Libraries may also be distributed as ZIP files that have the same name as the library; for example, the capacitive sensing library CapSense should be distributed as `CapSense.zip` and should contain a folder of the same name when unzipped.

Inside the folder there are files ending in `.h` and `.cpp`, such as `CapPin.h` and `CapPin.cpp`, and maybe even an Examples folder. If your `.zip` file contains only loose `.h` and `.cpp` files, you should place them in a folder with a library name. Sometimes you may find many `.h` and `.cpp` files that all perform different functions in the library, so make sure they're all inside the folder.

In the latest release of Arduino (1.8.5 at the time of this writing), it's easy to include libraries distributed as ZIP files. Simply choose Sketch⇨Include Library⇨Add .ZIP Library and select the ZIP file containing the library.

After the library is installed, restart Arduino and choose Sketch⇨Include Library to check that your library is in the list, as shown in Figure 12-5.

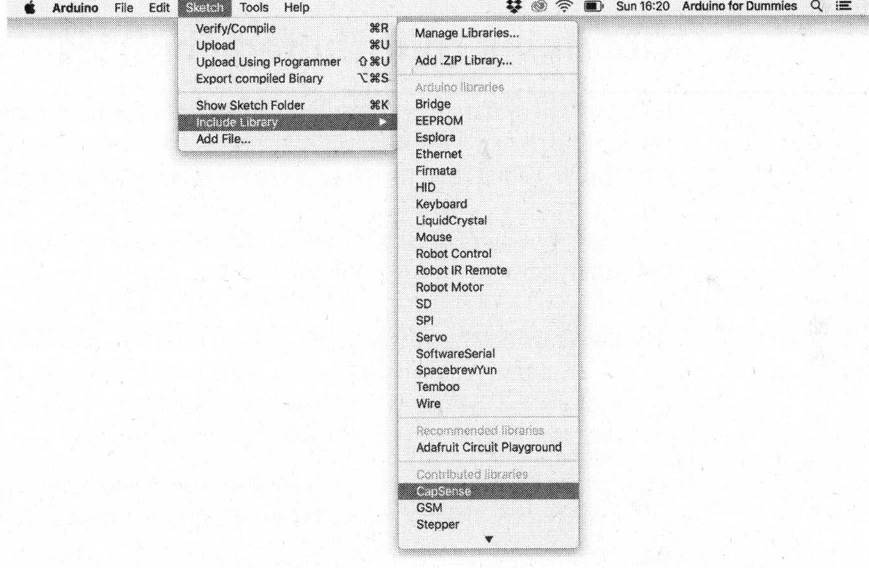

FIGURE 12-5: The Arduino menu shows the library in the Import Library drop-down list.

If the library has an Examples folder, you should also be able to see the examples by choosing File⇨Examples and then choosing the name of the library, as shown in Figure 12-6.

That's all there is to installing a library. Removing a library is as simple: just take the library folder out of the Arduino Sketch folder.

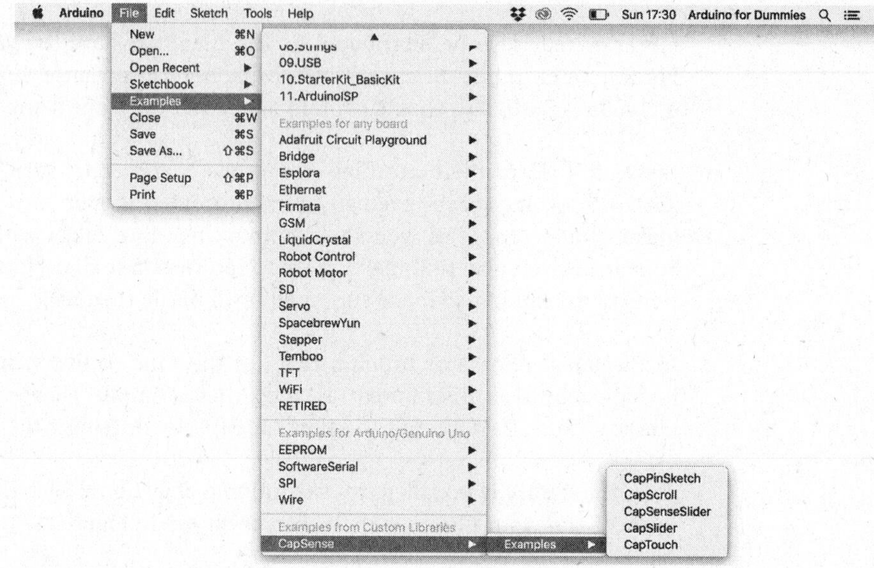

FIGURE 12-6: If there are examples with the library, you'll see them in the menu.

Obtaining contributed libraries

A long list of community-contributed libraries appears on the Arduino libraries page (`arduino.cc/en/Reference/Libraries`) and an exhaustive list appears on the Arduino Playground (`arduino.cc/playground/Main/LibraryList`).

To become familiar with contributed libraries, start with CapSense and TimerOne, two commonly used and helpful libraries:

- **CapSense (**`www.arduino.cc/playground/Main/CapSense`**):** The CapSense library allows you to make one or many pins on your Arduino into capacitive sensors. This feature allows you to make simple touch, pressure, or presence detection sensors quickly and easily with little hardware.

  The Arduino Playground page has a lot of useful information, but a more recent version of the code can be found on GitHub (`github.com/moderndevice/CapSense`).

- **TimerOne (**`playground.arduino.cc//Code/Timer1`**):** TimerOne (also called Timer1) uses a hardware timer on your Arduino to perform timed events at regular intervals. It's a great library for reading sensor data regularly without interrupting what's going on in the main loop. You can find a TimerOne page on the Arduino Playground and an up-to-date version of the library on Google Code (`code.google.com/p/arduino-timerone/`).

If you're keen to understand libraries more and maybe even write your own, check out the introduction to writing your own libraries on the Arduino GitHub page at `github.com/arduino/Arduino/wiki/Library-Manager-FAQ`.

4 Sussing Out Software

IN THIS PART . . .

Combine your knowledge of electronics with computer software to create stunning interactive visuals in the virtual world that exists on your computer.

Use data from software, your computer, or the Internet to create light, sound, and motion in the real world.

» Making shapes of all sizes and colors

Chapter 13
Getting to Know Processing

In the previous chapters, you learn all about using Arduino as a stand-alone device. A program is uploaded to the Arduino and carries out its task ad infinitum, until it is told to stop or powered down. You are affecting the Arduino by simple, clear electrical signals, and as long as no outside influences or coding errors exist and if the components last, the Arduino reliably repeats its function. This simplicity is useful for many applications and allows the Arduino to not only serve as a great prototyping platform but also work as a reliable tool for interactive products and installations for many years, as it already does in many museums.

Although this simplicity is something to admire, many applications are outside the scope of an Arduino's capabilities. Although the Arduino is basically a computer, it's not capable of running comparably large and complex computer programs in the same way as your desktop or laptop. Many of these programs are highly specialized depending on the task you're doing. You could benefit hugely if only you could link this software to the physical world in the same way your Arduino can.

Because the Arduino can connect to your computer and be monitored over the serial port, other programs may also be able to do this, in the same way that your computer talks to printers, scanners, or cameras. So by combining the physical world interaction capabilities of your Arduino with the data-crunching software capabilities of your computer, you can create projects with an enormous variety of inputs, outputs, and processes.

Many specific programs are made for specific tasks, but until you want to specify, it's best to find software that you can experiment with — that is, be a jack-of-all-trades in the same way that your Arduino is for the physical world. Processing is a great place to start.

In this chapter, you learn about Processing, the sister project that was in the first stages of development around the same time as Arduino. Processing is a software environment that you can use to sketch programs quickly, in the same way that you use an Arduino to test circuits quickly. Processing is a great piece of open source software, and its similarities to Arduino make it easy to learn.

Looking Under the Hood

An Arduino can communicate over its serial port as a serial device, which can be read by any program that can talk serial. Many programs are available, but Processing is one of the most popular.

Processing has an enormous breadth of applications ranging from visualizing data to creating generative artwork to performing motion capture using your webcam for digital performances. These are just a few niches; you can find a wealth of examples at `processing.org/exhibition`.

Processing is written in a Java-based language that looks similar to C (on which Arduino code is based) and C++. It is available for Windows, macOS, and Linux. Ben Fry and Casey Reas developed Processing to allow anyone, not just developers and engineers, to experiment with code. In the same way that ideas are sketched out, Processing is designed to sketch software. Programs can be quickly developed and adapted without a huge investment of time.

Processing uses a text-based IDE (integrated development environment) similar to that of Arduino. (In fact, it was "borrowed" by the Arduino team when the Arduino IDE was in development). A window displays the Java applet that the code creates, as shown in Figure 13-1. As with Arduino, the strength of Processing is the vast community that shares and comments on sketches, allowing the many participants to benefit from a diverse array of creative applications Processing is open source and allows users to modify the software as well as use it.

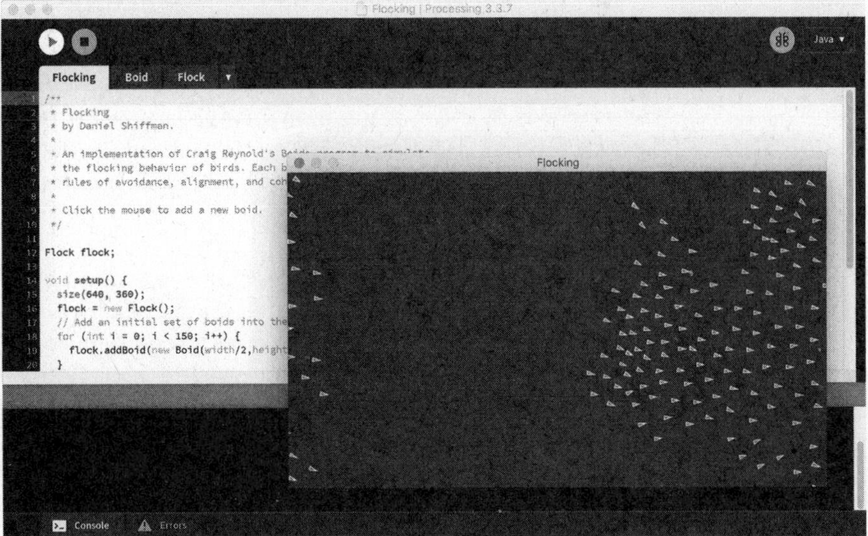

FIGURE 13-1: A typical view of Processing.

In this chapter, you learn how to get started with Processing. For more information, head over to the Processing site at `processing.org`.

TIP

Many other programming languages exist that can interface with Arduino. I describe Max/Pure Data and OpenFrameworks in sidebars in this chapter. For a list of even the most obscure languages, check the Arduino Playground at `arduino.cc/playground/main/interfacing`.

Installing Processing

Processing is free to download from `https://processing.org/download/` and supported on macOS, Windows 32-bit and 64-bit, and Linux 32-bit, 64-bit, and ARM. At the time of writing, Processing was version 3.3.7. Remember that things may have changed between when I wrote these words and when you get started.

To install Processing:

- **On a Mac:** The ZIP file unzips automatically, revealing the Processing app, which you can then drag to the Applications folder. From there you can drag Processing to the dock for easy access or create a desktop alias.
- **On Windows:** Unzip the ZIP file and place the Processing folder on your desktop or in a sensible location such as your Program Files folder: `C:/Program Files/Processing/`. Create a shortcut to `Processing.exe` and place it somewhere convenient, such as on your desktop or in the Start menu.

MAX/Pure Data

Max (also previously known as Max/MSP) is a visual programming language with a vast variety of uses, but it is most commonly used for audio, music, and sound synthesis applications. Max is available for Windows, macOS, Linux, and other more obscure operating systems.

Unlike traditional text-based programming languages, Max uses a graphical user interface to connect visual objects to one another in the same way that traditional synthesizers can be "patched" using wires to connect the various functions of the instrument. Software company Cycling '74 released the commercial software Max in 1990 based on earlier work by Miller Puckette to create a system for interactive computer music. Although the software is not open source, the application programming interface (API) allows third parties to make their own extensions to the software for specific uses. Miller Puckette also developed a redesigned, free, and open source version called PureData (Max).

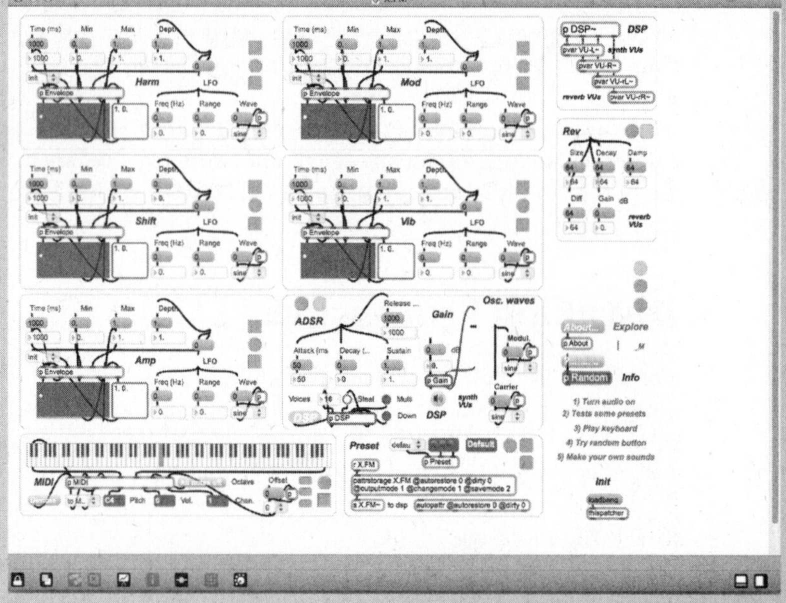

You can find more information about Max and PureData (Max) on their respective web pages: `cycling74.com/products/max` and `puredata.info`. To start communicating between Max and Arduino, check out the aptly named Maxuino (`www.maxuino.org/archives/category/updates`), and for PureData (Max) and Arduino, check out Pduino (`at.or.at/hans/pd/objects.html`).

Other helpful links are on the Arduino playground at (`www.arduino.cc/playground/interfacing/MaxMSP` and `www.arduino.cc/playground/interfacing/PD`).

Taking a look at Processing

After you have installed Processing, run the application. Processing opens with a blank sketch (see Figure 13-2) similar to the Arduino window, divided into five main areas:

» Toolbar with buttons

» Tabs

» Text editor

» Message area

» Console

The blank sketch also contains a menu bar for the main Processing application, which gives you drop-down menus to access the preferences of the processing application, load recent sketches and import libraries, and perform many other functions.

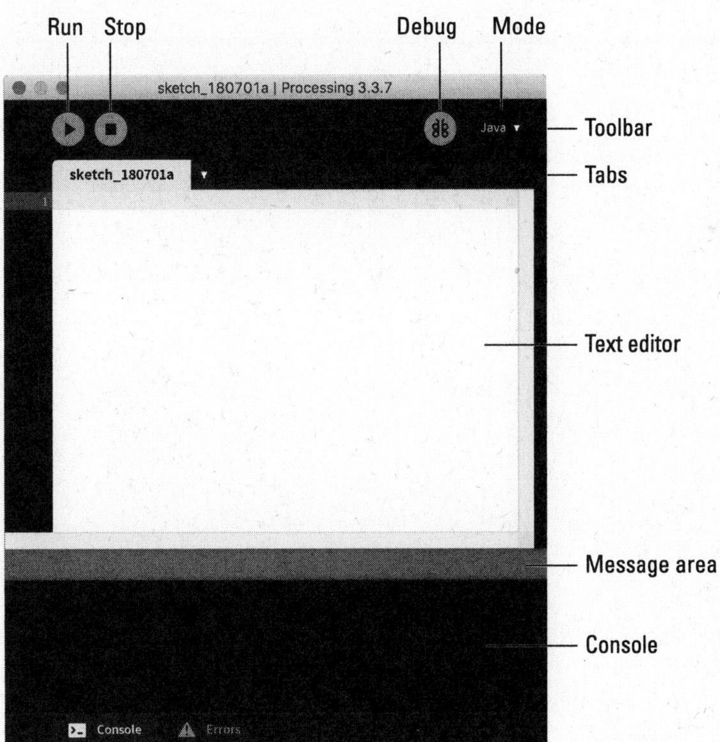

FIGURE 13-2: The Processing application is similar to but different from the Arduino one.

OPENFRAMEWORKS

OpenFrameworks, an open source C++ tool kit for experimenting with code, is actively developed by Zachary Lieberman, Theo Watson, and Arturo Castro, as well as other members of the openFrameworks community. OpenFrameworks runs on Windows, macOS, Linux, iOS, and Android. OpenFrameworks, unlike Processing, is not based on Java. Instead, OpenFrameworks is a C++ library designed to be the bare bones for getting started with audio-visual applications.

OpenFrameworks is especially powerful with graphics, allowing you to easily use OpenGL for intensive rendering or video applications. In contrast to Processing, Max, and PureData (Max), OpenFrameworks is not its own language; it is a collection of open source libraries, known as a *software framework* — hence the name. Because OpenFrameworks does not have its own IDE, the software used to write and compile the code depends on the platform. This feature can make getting started difficult because no centrally controlled IDE exists for continuity. The benefit is that C++ is highly versatile and can be used on almost any platform you can think of, including mobile operating systems.

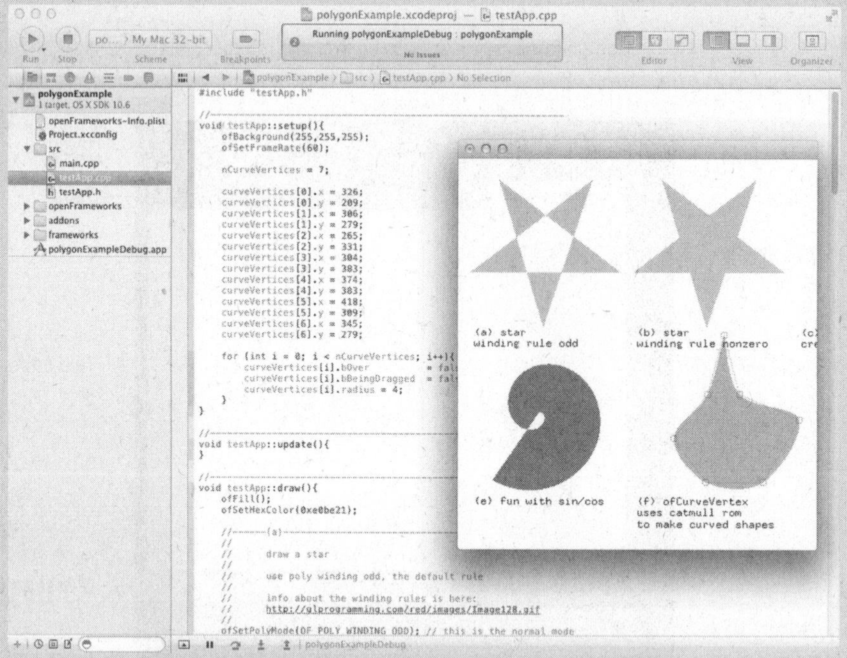

You can find details and tutorials at `www.openframeworks.cc/` and `www.openframeworks.cc/learning`. SparkFun also has a great Arduino tutorial for using OpenFrameworks with Arduino on Windows at `www.sparkfun.com/tutorials/318`.

Here's an overview of the Processing toolbar:

- **Run:** Executes or runs the code in the text editor as an applet (small application) in a new window. The keyboard shortcuts for this command are Ctrl+R for Windows and Cmd+R for macOS.
- **Stop:** Stops the code from running and closes the applet window.
- **Debug:** A tool to help you identify errors and debug your code.
- **Mode:** Changes mode between Java (standard), Android (mobile and tables), and JavaScript (online applications). This capability is a new development in the latest release. You can find more details on these modes at `github.com/processing/processing-android/wiki` and `github.com/fjenett/javascript-mode-processing/wiki`.
- **Tabs:** Organizes multiple files in a Processing sketch. Use tabs in larger programs to separate objects from the main sketch or to incorporate look-up tables of data into a sketch.
- **Text editor:** Enters code into the sketch. Recognized terms or functions are highlighted in appropriate colors for clarity. The text editor is the same as that in the Arduino IDE.
- **Message area:** Displays errors, feedback, or information about the current task. You might see a notification that the sketch saved successfully, but more often than not, the message shows where errors are flagged.
- **Console:** Displays more details on your sketch. You can use the `println()` function here to display the values in your sketch; additional detail on errors is also shown.

Trying Your First Processing Sketch

Unlike with Arduino, you don't need an extra kit to get going with Processing. This feature makes Processing useful for learning about coding because you can enter a line or two of code, click Run, and see what the results.

Start your first sketch with these steps:

1. **Press Ctrl+N (in Windows) or Cmd+N (on a Mac) to open a new sketch.**
2. **Click in the text editor and enter this line of code:**

```
ellipse(50,50,10,10);
```

3. **Click the Run button.**

   A new applet window opens, showing a white circle in the middle of a gray box, as in Figure 13-3.

Well done! You've just written your first Processing program.

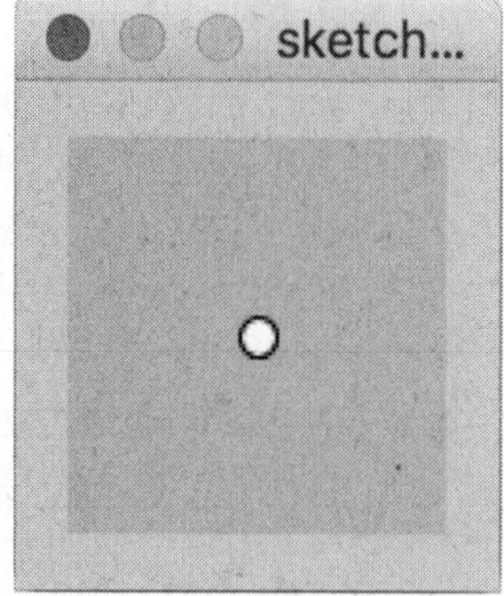

FIGURE 13-3: A Processing sketch that draws an ellipse with equal dimensions, also known as a circle.

Have you finished admiring your circle? That line of code draws an ellipse. An ellipse normally is not circular, but you gave it the parameters to make a circle.

The word *ellipse* is highlighted in turquoise in the text editor, indicating that it is a recognized function. The first two numbers are the coordinates of the ellipse, which in this case are 50, 50. The unit of the numbers is in pixels. Because the default window is 100 x 100 pixels, coordinates of 50, 50 put the ellipse in the center. The 10, 10 values indicate the width and height of the ellipse, giving you a circle. You could write the function also as

```
ellipse(x,y,width,height)
```

The coordinates for the ellipse (or any shape or point, for that matter) are written as x and y. These indicate a point in two-dimensional (2D) space, which in this case is a point measured in pixels on your screen. Horizontal positions are referred to as the x coordinate; vertical positions are the y coordinate. Depth used in 3D space is referred to as z. Add the following line of code, just above the `ellipse()` statement:

```
size(300,200);
```

Click the Run button and you get a rectangular window with the ellipse in the top left, as shown in Figure 13-4. The `size()` function is used to define the size of the applet window in pixels, which in this case is 300 pixels wide and 200 pixels high. If your screen isn't like Figure 13-4, you may have put the statements in the wrong order. The lines of code are read in order, so if the ellipse code is first, the blank window is drawn over the ellipse. And with a rectangular window, you see that the coordinates are measured from the top left.

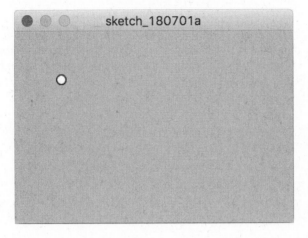

FIGURE 13-4: A resized display window.

Coordinates are measured on an invisible a grid with the center point at 0, 0 for 2D (or 0, 0, 0 for 3D), which is referred to as the *origin*. This way of referencing locations is based on the Cartesian coordinate system, which you may have studied in school. Numbers can be positive or negative, depending on which side of the origin they are on. On computer screens, the origin is at the top left because pixels are drawn from top left to bottom right, one row at a time (check out Figure 13-5). Therefore, the statement `size(300,200)` draws a window 300 pixels from left to right on the screen and then 200 pixels from top to bottom.

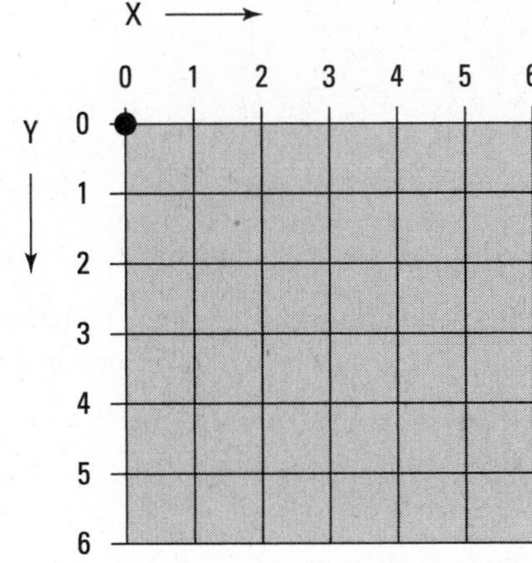

FIGURE 13-5: How the grid looks on a computer.

Drawing shapes

To gain a better understanding of the possibilities you have in drawing shapes, look at a few basic shapes: point, line, rectangle, and ellipse.

Point

A single point is the most basic shape and is useful for creating more complex shapes. Write the following code and then click the Run button. Look closely and you'll see a single black pixel in the center of the display window (see Figure 13-6). That is the point that your code drew.

```
size(300,200);
point(150,100);
```

The `point` function can also be written as follows:

```
point(x,y);
```

FIGURE 13-6: If you look closely, you can see the point.

Line

A line is made by connecting two points, which is done by defining the start and end points. Write the code to generate a screen like the one in Figure 13-7:

```
size(300,200);
line(50,50,250,150);
```

You can also write a line written as follows:

```
line(x1,y1,x2,y2);
```

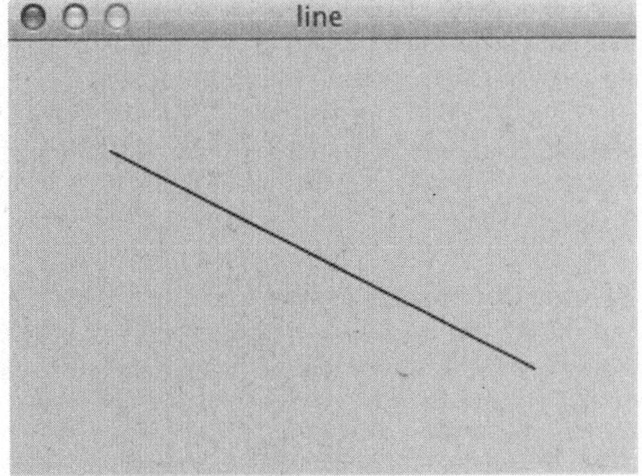

FIGURE 13-7: A line between two points.

Rectangle

You can draw a rectangle a number of different ways. In this first example, a rectangle is drawn by identifying the starting point and then the width and height of the rectangle. Write the following code to draw a rectangle in the center of your display window:

```
size(300,200);
rect(150,100,50,50);
```

In this case, you have a rectangle that starts in at point 150,100 in the center of the display window. That is the top-left corner of the rectangle, and from there it has a width of 50, which extends the rectangle toward the right of the window, and a height of 50, which extends toward the bottom of the window. This function is particularly useful if you want the size of the rectangle to remain constant but change the position of the rectangle. You could also write the code as follows:

```
rect(x,y,width,height);
```

When drawing rectangles, you can choose among different modes (see Figure 13-8). If the mode is set to center, the rectangle is drawn centered around a point instead of being drawn from that point. Write the following code and you see that the same values display a different position when `rectMode` is changed to `CENTER`:

```
rectMode(CENTER);
size(300,200);
rect(150,100,50,50);
```

You can see that the rectangle is now centered in the display window. The shape extends equally from the center point both left to right and top to bottom.

You can also draw a rectangle by declaring two diagonally opposite corners. Write the following code, this time with `rectMode` set to `CORNERS`:

```
rectMode(CORNERS);
size(300,200);
rect(150,100,50,50);
```

You see a rectangle that is quite different from the others because it starts at the same point in the center, 150,100, but ends at point 50,50, creating a rectangle positioned above and to the left of the starting point. You can also write the code as follows:

```
rect(x1,y1,x2,y2);
```

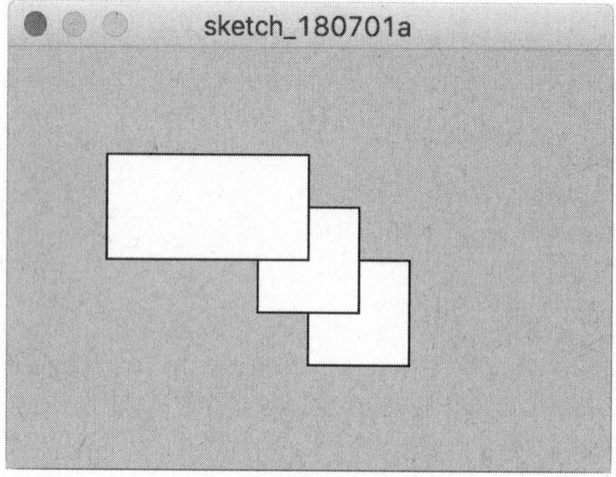

FIGURE 13-8: A selection of differently drawn rectangles.

Ellipse

The first item covered in this chapter was `ellipse`, which can be used to simply draw an ellipse. Write the following code to draw an ellipse in the center of the display window:

```
ellipse(150,100,50,50);
```

The default mode for `ellipse` is CENTER, whereas the default mode for `rect` is `CORNER`. The preceding line of code can also be written as follows:

```
ellipse(x,y,width,height);
```

As with `rectMode()` it's possible to set different modes (see Figure 13-9) for drawing ellipses using `ellipseMode()`. Write the following code to draw an ellipse from its corner instead of its center:

```
ellipseMode(CORNER);
size(300,200);
ellipse(150,100,50,50);
```

This draws an ellipse from starting from its top-left corner with a width of 50 and a height of 50.

It is also possible to draw an ellipse by specifying multiple corners. Write the following code to change the `ellipseMode` to `CORNERS`:

```
ellipseMode(CORNERS);
size(300,200);
ellipse(150,100,50,50);
```

Similarly to `rectMode(CORNERS)`, the `ellipseMode(CORNERS)` creates an ellipse positioned above and to the left of the starting point. The first corner is the center point of the sketch and the second is at point 50,50.

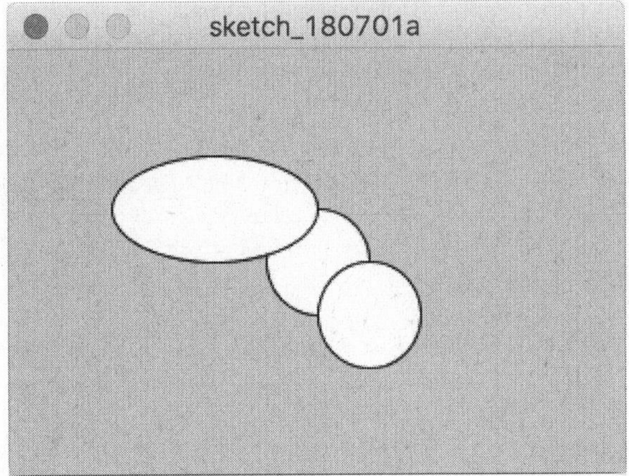

FIGURE 13-9: A selection of differently drawn ellipses.

Changing color and opacity

Now that you have an understanding of shapes, it's time to affect their appearance. The simplest way to change a shape's appearance is to change the color. By using the `background` and `fill` functions, you can change the color of the shapes on your screen to any one of 16,777,216 colors. You can also set the opacity of the objects you draw, allowing you to mix colors by layering semitransparent shapes.

MAKING SENSE OF BINARY, BITS, AND BYTES

The binary number system, also known as base-2, uses only two values: 0 or 1. Decimal numbers use 0 to 9 and are usually referred to as base-10. Hexadecimal numbers use 0 to 9 and A to F and are referred to as base-16.

But how is binary useful when you're trying to talk to lots of things and have only two options? The answer is that you use a lot of binary values.

For example, if you have a base-2 binary number such as 10101101, you can determine its value in base-10 with a simple lookup table. Binary is typically read from right to left. Because binary is base-2, each value is the binary value multiplied by 2 to the power of (2^x), where *x* is equal to the order of the bit, starting at 0 on the right. For example, as shown in the following, the fourth binary value is equal to 1x(2x2x2) = 8.

Binary	1	0	1	0	1	1	0	1	
Calculation	1×2^7	0×2^6	1×2^5	0×2^4	1×2^3	1×2^2	0×2^1	1×2^0	Total
Decimal	128	0	32	0	8	4	0	1	173

As you can see, extremely large numbers can be formed using only zeros and ones. In this case, you have eight binary values with a total decimal value of 255. When talking about memory, each binary value takes one *bit* of memory, and each group of eight bits is referred to as a *byte*. To give you an idea of scale, a blank Arduino sketch uses 466 bytes; an Uno can store a maximum of 32,256 bytes, and a Mega can store a maximum of 258,048 bytes.

Background

The `background` function changes the background of your sketch. You can choose grayscale values or color. You'll start by changing grayscale values.

Open a new sketch, and type the following code to change the default gray window to black:

```
background(0);
```

Change 0 to 255 to change the color to white:

```
background(255);
```

Any value between 0 (black) and 255 (white) is a grayscale value. The reason that this range is 0 to 255 is that there are 8 bits of data in a byte, meaning that you need one byte to store a grayscale color value.

To liven things up a bit, you can add color to your sketch background. Instead of 8-bit grayscale, you use 24-bit color, which is 8-bit red, 8-bit green, and 8-bit blue. The color of the background is defined with three values instead of one:

```
background(200,100,0);
```

This line of code gives you an orange background, which is comprised of a red value of 200, a green value of 100, and a blue value of 0. There are several color modes, but in this case this line of code can be interpreted as follows:

```
background(red,green,blue);
```

Fill

Want to change the color of the shapes you draw? Use `fill` to both set color and control the shape's opacity.

`fill` sets the color for any shape that is drawn after it. By calling `fill` multiple times, you can change the color of several different shapes. Write the following code to draw three ellipses with different colors, as shown in Figure 13-10:

```
background(255);
noStroke();

// Bright red
fill(255,0,0);
ellipse(50,35,40,40);

// Bright green
fill(0,255,0);
ellipse(38,55,40,40);

// Bright blue
fill(0,0,255);
ellipse(62,55,40,40);
```

The background is set to white (255), and the `noStroke` function removes borders from the shapes. You can comment out the function by using two forward slashes (//), to see the effect.

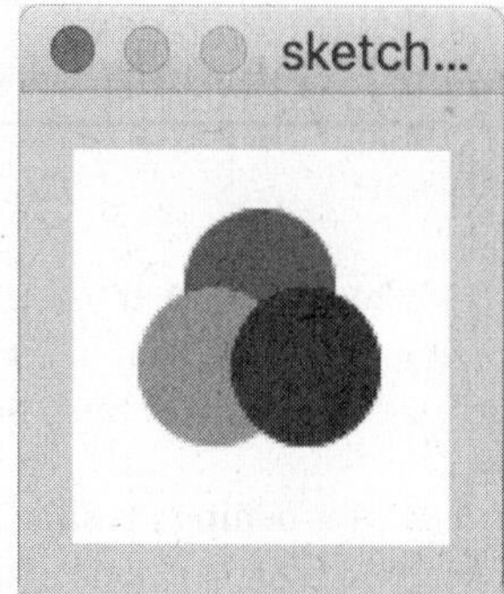

FIGURE 13-10: The different colored circles.

It's also important to note that all shapes are drawn in the order in which they were programmed. You can see that the first circle to be drawn is red because the other two circles are layered on top of it. The red value is the highest possible (255), as is the second for green and the third for blue. If another shape were drawn at the end of the code, it would be the same strong blue because that is the last `fill` value.

You can also affect the opacity of the colors, creating semitransparent shapes. By adding a fourth value to the `fill` function, you can set the opacity from 0 (fully transparent) to 255 (solid color). Update the preceding code with the following values to give the circles transparency:

```
background(255);
noStroke();

// Bright red
fill(255,0,0,100);
ellipse(50,35,40,40);

// Bright green
fill(0,255,0,100);
ellipse(38,55,40,40);

// Bright blue
fill(0,0,255,100);
ellipse(62,55,40,40);
```

Playing with interaction

All of this is fun but static. In the next example, you inject some life into your sketches by using your mouse as an input. To do this, you must constantly update the sketch by looping through it over and over again, sending new values for each loop. Write the following code to create an interactive sketch:

```
void setup() {
}

void draw() {
ellipse(mouseX,mouseY,20,20);
}
```

This code draws an ellipse centered on your mouse pointer coordinates, so when you move your mouse you leave a trail of ellipses behind, as shown in Figure 13-11. The functions `mouseX` and `mouseY` are shown in pink in the text editor and take the coordinates of your mouse pointer in the display window. The values are the number of pixels horizontally and vertically, respectively.

This code might look familiar. Instead of Arduino's `void setup` and `void loop`, Processing uses `void setup` and `void draw`. These work in almost the same way: `setup` runs once at the start of the sketch; `loop` and `draw` run forever or until they are told to stop. You can stop the sketch at any time by clicking the Stop button.

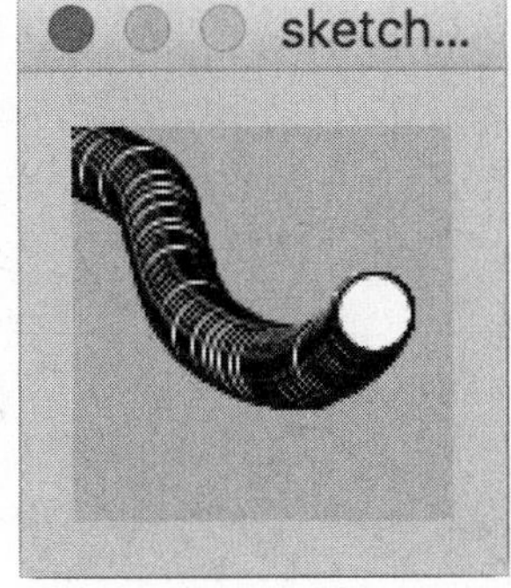

FIGURE 13-11: Drawing lots of ellipses wherever your mouse pointer goes.

Change the sketch slightly, and you can cover up all those previous ellipses to display only the most recent (see Figure 13-12).

```
void setup() {
}

void draw() {
background(0);
ellipse(mouseX,mouseY,20,20);
}
```

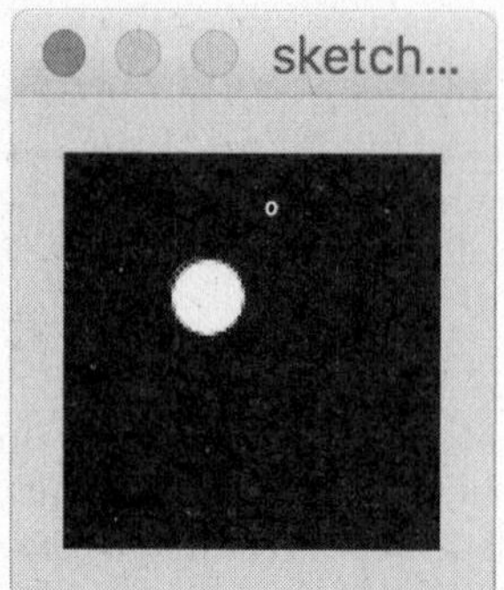

FIGURE 13-12: Just one ellipse at the location of your mouse pointer.

There is much more to Processing that I can't cover in this book. However, these few points should be enough to gain a basic comprehension of how code relates to the onscreen visuals. You can find a wealth of examples, both on the Processing site and in the Processing software. The best approach is to run the examples and then tweak the values to see what happens. You'll learn what's going on much more quickly by experimenting, and with no electronics, you won't break anything.

» Graphing data from the physical world

» Sending signals between the Arduino and Processing

Chapter 14

Processing the Physical World

In the preceding chapter, you learn the basics of Processing and its similarities to and differences from Arduino. This chapter is all about combining both tools to integrate the virtual and physical worlds. These few exercises teach you the basics about sending and receiving data in both Processing and Arduino. You can build on this knowledge to create your own projects, maybe to generate some awesome onscreen visuals from your sensors or to turn on a light every time someone mentions you on Twitter.

Making a Virtual Button

In this example, you learn how to make an onscreen button in Processing that affects a physical LED on your Arduino. This is a great sketch to get started with interactions between computers and the real world, and between an Arduino and Processing.

You need the following:

- An Arduino Uno
- An LED

The setup is simple for this introduction to Arduino and Processing, requiring only a single LED.

As shown in Figures 14-1 and 14-2, insert the long leg of the LED into pin 13 and the short leg into GND. If you don't have an LED, you can simply monitor the onboard LED marked L.

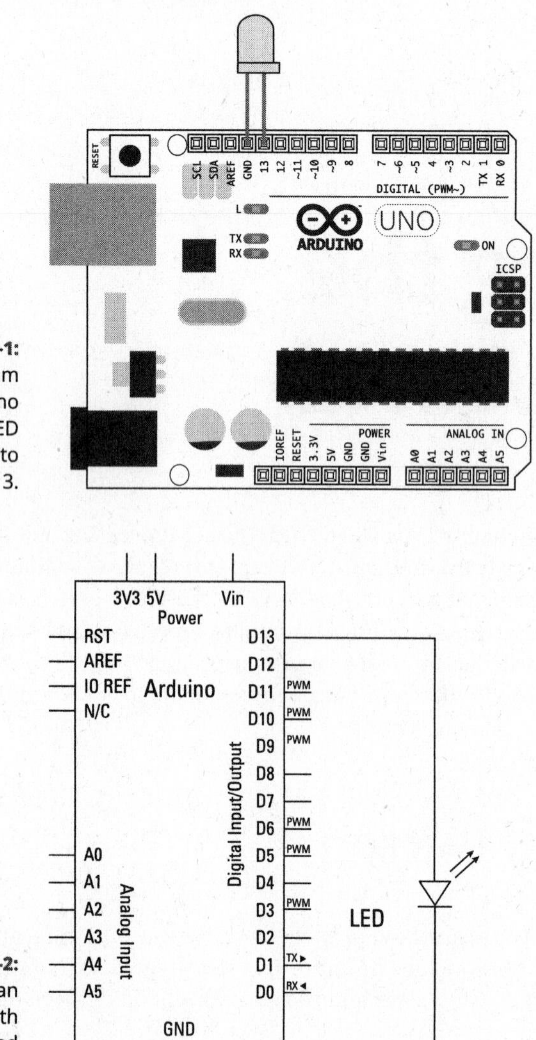

FIGURE 14-1: A circuit diagram of an Arduino with an LED connected to pin 13.

FIGURE 14-2: A schematic of an Arduino with an LED connected to pin 13.

Setting up the Arduino code

After your circuit is assembled, you need the appropriate software to use it. From the Arduino menu, choose File ➪ Examples ➪ 04.Communication ➪ PhysicalPixel to open the sketch.

This sketch contains both Arduino code and the relevant Processing code for the sketch to work. (It also has a variation in Max 5.) The code below the Arduino code is commented out to avoid interfering with the Arduino code.

TECHNICAL STUFF

In older versions of Arduino, the sketch files ended with .pde, which is the Processing suffix. This caused confusion, so now the Arduino suffix is .ino. Different suffixes make it possible to have the Arduino sketch and the Processing sketch in the same place. If you try to open a .pde file in Arduino, the application assumes that it is an old Arduino sketch and asks whether you want to change the suffix to .ino.

```
/*
  Physical Pixel

 An example of using the Arduino board to receive data from the
 computer.  In this case, the Arduino boards turns on an LED when
 it receives the character 'H', and turns off the LED when it
 receives the character 'L'.

 The data can be sent from the Arduino serial monitor, or another
 program like Processing (see code below), Flash (via a serial-net
 proxy), PD, or Max/MSP.

 The circuit:
 * LED connected from digital pin 13 to ground

 created 2006
 by David A. Mellis
 modified 30 Aug 2011
 by Tom Igoe and Scott Fitzgerald

 This example code is in the public domain.

 http://www.arduino.cc/en/Tutorial/PhysicalPixel
 */

const int ledPin = 13; // the pin that the LED is attached to
int incomingByte;      // a variable to read incoming serial data into
```

```
void setup() {
  // initialize serial communication:
  Serial.begin(9600);
  // initialize the LED pin as an output:
  pinMode(ledPin, OUTPUT);
}

void loop() {
  // see if there's incoming serial data:
  if (Serial.available() > 0) {
    // read the oldest byte in the serial buffer:
    incomingByte = Serial.read();
    // if it's a capital H (ASCII 72), turn on the LED:
    if (incomingByte == 'H') {
      digitalWrite(ledPin, HIGH);
    }
    // if it's an L (ASCII 76) turn off the LED:
    if (incomingByte == 'L') {
      digitalWrite(ledPin, LOW);
    }
  }
}
```

Now go through the steps to upload your sketch.

With the Arduino set up to receive a message from Processing, you need to set up the Processing sketch to send a signal message over the same serial port to your Arduino.

Setting up the Processing code

The Processing code is available within multiline comment markers (`/* */`) at the bottom of the Arduino PhysicalPixel sketch. Copy the code within the comment markers, paste it into a new Processing sketch, and save it with an appropriate name, such as PhysicalPixel:

```
// Mouse over serial

 // Demonstrates how to send data to the Arduino I/O board, in order to
 // turn ON a light if the mouse is over a square and turn it off
 // if the mouse is not.

 // created 2003-4
 // based on examples by Casey Reas and Hernando Barragan
 // modified 30 Aug 2011
```

```
// by Tom Igoe
// This example code is in the public domain.



import processing.serial.*;

float boxX;
float boxY;
int boxSize = 20;
boolean mouseOverBox = false;

Serial port;

void setup()  {
size(200, 200);
boxX = width/2.0;
boxY = height/2.0;
rectMode(RADIUS);

// List all the available serial ports in the output pane.
// You will need to choose the port that the Arduino board is
// connected to from this list. The first port in the list is
// port #0 and the third port in the list is port #2.
println(Serial.list());

// Open the port that the Arduino board is connected to (in this case #0)
// Make sure to open the port at the same speed Arduino is using (9600bps)
port = new Serial(this, Serial.list()[0], 9600);

}

void draw()
{
background(0);

// Test if the cursor is over the box
if (mouseX > boxX-boxSize && mouseX < boxX+boxSize &&
mouseY > boxY-boxSize && mouseY < boxY+boxSize) {
mouseOverBox = true;
// draw a line around the box and change its color:
stroke(255);
fill(153);
// send an 'H' to indicate mouse is over square:
port.write('H');
}
```

```
else {
// return the box to its inactive state:
stroke(153);
fill(153);
// send an 'L' to turn the LED off:
port.write('L');
mouseOverBox = false;
}

// Draw the box
rect(boxX, boxY, boxSize, boxSize);
}
```

Click the Run button to execute the Processing sketch, and an applet appears. The applet has a black background with a gray square in the middle, representing your virtual button (shown in Figure 14-3). If you move your mouse cursor over the gray square (or pixel), you can see that its edges turn white. If you then look at your Arduino, you see that whenever your mouse cursor hovers over the gray square, the LED on your board illuminates, giving you a physical representation of your pixel.

FIGURE 14-3: Your Processing applet displaying the virtual pixel.

If your LED doesn't light, double-check your wiring:

- Make sure that you're using the correct pin number.
- Make sure that the LED legs are inserted correctly.
- Check that you're using the correct serial port.
- Check that your Arduino code uploaded correctly and that your Processing code has no errors. Note that you cannot upload while the Processing sketch is communicating with your Arduino, so you need to stop the sketch before uploading.

Understanding the Processing PhysicalPixel sketch

Wherever possible, dividing projects into their elements is a good idea. You may have many inputs and outputs, but if you deal with them one at a time, they are easier to understand and easier to troubleshoot. Because the Processing side is the input, you're wise to start with that.

The structure of a Processing sketch is similar to Arduino. You include libraries and declare variables at the start of the sketch and set fixed values or initializations in `setup`. The `draw` function then repeats its process until told otherwise.

Processing uses libraries to add functionality in the same way as Arduino does. In this case, a serial communication library is needed to talk to Arduino. In Arduino, this library is included by using `#include <libraryName.h>`. However, in Processing, you use the `import` keyword, followed by the name and the syntax `*` to load all the related parts of that library:

```
import processing.serial.*;
```

A *float* is a floating-point number, one with a decimal place, such as 0.5 or 10.9. In this case, two floating-point numbers are declared, `boxX` and `boxY`. These are the coordinates for the location of the box:

```
float boxX;
float boxY;
```

Next, `boxSize` defines the size of the box as an integer, or whole number. Because it is square, only one value is needed:

```
int boxSize = 20;
```

A Boolean (which can be only `true` or `false`) is used to communicate that the mouse cursor is over the box. `boolean` is set to start as `false`:

```
boolean mouseOverBox = false;
```

The last thing to do is create a new `Serial port` object. Many serial connections could be in use by your computer, so it's important that each one be named so that it can be used as needed. In this case, you are using only one port. The word *serial* is specific to the serial library to indicate that you want to create a new serial object (connection), and the word *port* is the name of the object (connection) used to refer to the port from this point on. Think of it as giving your cat a collar. If too many cats are in a room, they all look fairly similar and follow the same general rules, but they are still all individual. If you put a colored collar on each cat with its name printed on it, you can easily identify which one is which.

```
Serial port;
```

In `setup`, the first item to define is the size of the display window, which is set to 200 pixels square:

```
void setup() {
  size(200,200);
```

The variables for `boxX` and `boxY` are set to be proportional to the width and height, respectively, of the display window. They are always equal to half the width and height. Next, `rectMode` is set to `RADIUS`, which is similar to `CENTER`, but instead of specifying the overall width and height of the rectangle, `RADIUS` specifies half the height and width. (`CENTER` could be interpreted as diameter in that respect.) Because the coordinates of the box are centered and are aligned to the center point of the display window, the box is also perfectly centered:

```
  boxX = width/2.0;
  boxY = height/2.0;
  rectMode(RADIUS);
```

Your computer may have a lot of serial connections, so it's best to print a list of them to locate your Arduino:

```
  println(Serial.list());
```

The most recent port usually appears at the top of this list in position 0, so if you've just plugged in your Arduino, the first item is likely the one you want. If you're not using the `Serial.list` function, you could replace `Serial.list()[0]` with another number in the list, which will be printed on the console. You can also replace `Serial.list()[0]` with the name of the port, such as /dev/tty.usbmodem26221 or COM5. Specifying the name is also useful if you have multiple Arduinos connected to the same computer. The number 9600 refers to the baud rate, which is the rate at which you're communicating with the Arduino.

```
port = new Serial(this, Serial.list()[0], 9600);
}
```

If the baud rate number is not the same on both ends (the sending and receiving end), the data will not be received.

In draw, the first task is to draw a black background:

```
void draw()
{
  background(0);
```

Processing uses the same (or similar) conditionals as Arduino. This if statement tests the mouse value to see whether the cursor is over the box area. If mouseX is greater than the box coordinate (center) minus the size of the box (half the box width), and less than the box coordinate (center) plus the size of the box (half the box width), the horizontal position is over the box. This statement is used again with the vertical position, using AND statements (&&) to add to the conditions of the if statement. Only if all these are true can the Boolean mouseOverBox be declared true:

```
  // Test if the cursor is over the box
  if (mouseX > boxX-boxSize && mouseX < boxX+boxSize &&
    mouseY > boxY-boxSize && mouseY < boxY+boxSize) {
    mouseOverBox = true;
```

To indicate that mouseOverBox is true, the code draws a white line around the box. Rather than requiring that another box be drawn, the white line appears simply by changing the stroke, or outline, value (*stroke* is a term common in most graphics software). The stroke is set to 255, which outlines the box in white:

```
    // draw a line around the box and change its color:
    stroke(255);
```

Fill is set to 153, a mid-gray, which colors the next object drawn:

```
    fill(153);
```

Then the all-important communication is sent. The port.write statement is similar to Serial.print but is used for writing to a serial port in Processing. The character sent is H, for high:

```
    // send an 'H' to indicate mouse is over square:
    port.write('H');
  }
```

The `else` statement tells Processing what to do if the mouse cursor is not over the box:

```
else {
```

The `stroke` value is set to the same mid-gray as the box. The box fill color remains the same whether active or inactive:

```
  // return the box to its inactive state:
  stroke(153);
  fill(153);
```

The character L is sent to the serial port to signify that the LED should be set low:

```
  // send an 'L' to turn the LED off:
  port.write('L');
```

The Boolean `mouseOverBox` is set to `false`:

```
  mouseOverBox = false;
}
```

Finally, the box (technically a rectangle) is drawn. Its coordinates are always centered, and its size remains the same; the only difference is the color applied by the `if` statement. If the mouse cursor is over the box, the `stroke` value is changed to white (active), and if not, the `stroke` value is set to the same gray as the box and appears to not be there (inactive):

```
  // Draw the box
  rect(boxX, boxY, boxSize, boxSize);
}
```

Understanding the Arduino PhysicalPixel sketch

In the preceding section, you find out how the Processing side provides a signal. The signal is sent over the serial connection to your Arduino. In this section, I explain what the Arduino code does with the signal. The Arduino code for this example is simple compared to other examples in this book and is great for understanding how a serial connection is made. I always recommend starting with this sketch for any Processing-to-Arduino communication. It's great as a foundation to make sure that your hardware and software are working, and you can then build on or adapt it as needed.

First, the constant and variable values are declared. The LED pin — pin 13 — is the LED output and does not change, so it is declared as a constant. The `incomingByte` value does change and is declared as an integer (`int`), not a character (`char`). I explain why a bit later.

```
const int ledPin = 13; // the pin that the LED is attached to
int incomingByte;      // a variable to read incoming serial data into
```

In `setup`, the serial communication is initialized and set to a matching baud rate of 9600.

REMEMBER

Remember that in Processing and Arduino, if you change the speed of the device or application sending data, you also must change the speed of the device or application receiving data. When communicating with a computer, you must choose from the following values: 300, 600, 1200, 2400, 4800, 9600, 14400, 19200, 28800, 38400, 57600, or 115200.

```
void setup() {
  // initialize serial communication:
  Serial.begin(9600);
```

Pin 13, or `ledPin` as it is named, is set to be an output.

```
  // initialize the LED pin as an output:
  pinMode(ledPin, OUTPUT);
}
```

The first action in the `loop` is to determine whether any data is available. `Serial.available` reads the serial buffer, which stores any data sent to the Arduino before it is read. Nothing happens until data is sent to the buffer.

By checking that the value is greater than 0, you reduce the number of readings considerably. Reading lots of 0 or null values can considerably slow down the operation of your Arduino and any programs or hardware reading from it:

```
void loop() {
  // see if there's incoming serial data:
  if (Serial.available() > 0) {
```

If a value is greater than 0, it is stored in the `int` variable `incomingByte`:

```
    // read the oldest byte in the serial buffer:
    incomingByte = Serial.read();
```

Now you need to know if the data received is what your program is expecting. Processing sent H as a character, but that is just a byte of data that can be understood as a number or a character. In this case, you're treating it as an integer. This `if` statement checks to see whether the integer value is equal to 72, which is equal to the character H in ASCII. The inverted commas indicate that it is a character and not a variable. The statement `if (incomingByte == 72) {` would return the same result:

```
// if it's a capital H (ASCII 72), turn on the LED:
if (incomingByte == 'H') {
```

If the values are equal, pin 13 is set `HIGH`:

```
  digitalWrite(ledPin, HIGH);
}
```

If the value is the character L, or the integer value 76, the same pin is set `LOW`:

```
  // if it's an L (ASCII 76) turn off the LED:
  if (incomingByte == 'L') {
    digitalWrite(ledPin, LOW);
  }
 }
}
```

This basic Processing-to-Arduino interaction works great as the basis for larger projects. In this example, the onscreen interaction is the input and could be swapped for more useful or elaborate inputs. One such input is face tracking: When your face is at the center of the screen, the signal is sent. On the Arduino side of the code, as is true of Processing, a vast array of outputs could be triggered besides lighting an LED (lovely though it is). For instance, you could link optocouplers to a remote and begin playback whenever a high signal is sent, and you could pause playback whenever a low signal is sent. See the bonus chapter, "Hacking Other Hardware," at `www.dummies.com/go/arduinofd` for more about working with optocouplers.

Drawing a Graph

In the preceding section, you see how to send a signal in one direction. Want to learn how to send signals the other direction, from Arduino to Processing? In this example, you find out how to read the value of a potentiometer using your Arduino and display it in a Processing applet.

You need the following:

- An Arduino Uno
- A breadboard
- A potentiometer
- Jump wires

This basic circuit uses a potentiometer to send an analog value to Processing that can be interpreted and displayed on an onscreen graph. Assemble the circuit by connecting the center pin of the potentiometer to analog pin 0, following Figures 14-4 and 14-5. The potentiometer is wired with the central pin connected to analog pin 0. Of the other two pins, one is connected to 5V and the other to GND. By reversing these pins, you can change the directions that the potentiometer counts when turned.

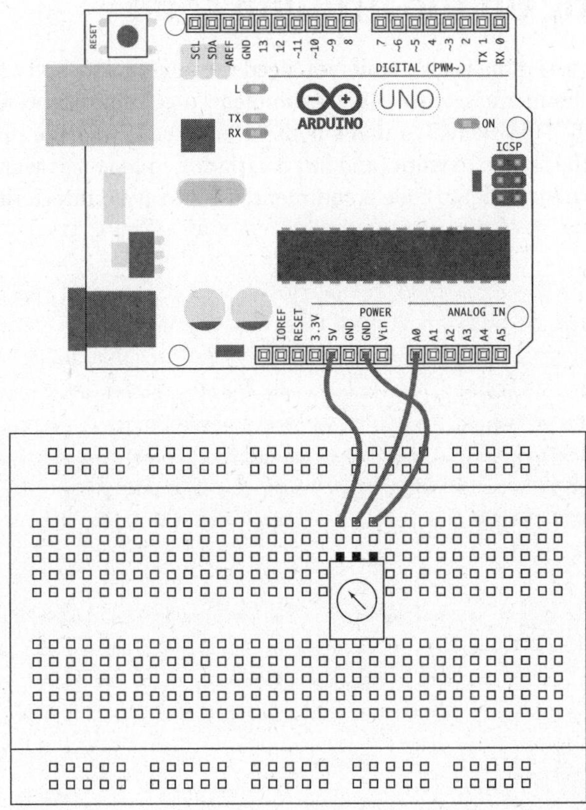

FIGURE 14-4: A circuit diagram for a potentiometer input.

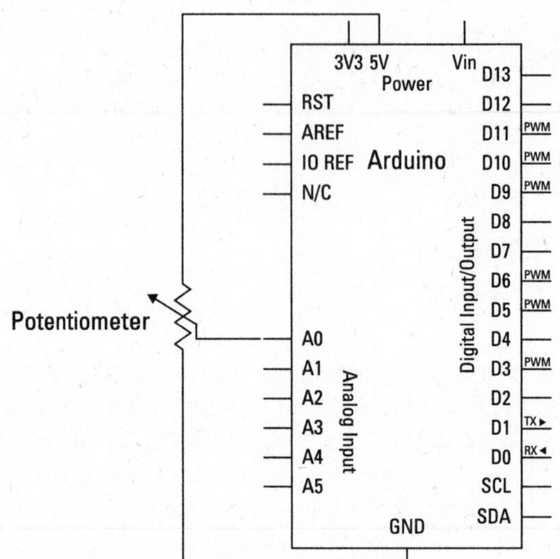

FIGURE 14-5: A schematic for a potentiometer input.

Setting up the Arduino code

After you assemble your circuit, you need the appropriate software to use it. From the Arduino menu, choose File ⇨ Examples ⇨ 04.Communication ⇨ Graph to open the sketch. This sketch contains both Arduino code and the relevant Processing code for the sketch to work (and has a variation in Max 5 as well). The Processing code below the Arduino code is commented out to avoid interfering with the Arduino code.

```
/*
  Graph

 A simple example of communication from the Arduino board to the computer:
 the value of analog input 0 is sent out the serial port.  We call this "serial"
 communication because the connection appears to both the Arduino and the
 computer as a serial port, even though it may actually use
 a USB cable. Bytes are sent one after another (serially) from the Arduino
 to the computer.

 You can use the Arduino Serial Monitor to view the sent data, or it can
 be read by Processing, PD, Max/MSP, or any other program capable of reading
 data from a serial port.  The Processing code below graphs the data received
 so you can see the value of the analog input changing over time.

 The circuit:
 - any analog input sensor attached to analog in pin 0.
created 2006
```

```
  by David A. Mellis
  modified 9 Apr 2012
  by Tom Igoe and Scott Fitzgerald

  This example code is in the public domain.

  http://www.arduino.cc/en/Tutorial/Graph
  */

void setup() {
  // initialize the serial communication:
  Serial.begin(9600);
}

void loop() {
  // send the value of analog input 0:
  Serial.println(analogRead(A0));
  // wait a bit for the analog-to-digital converter
  // to stabilize after the last reading:
  delay(2);
}
```

Now go through the steps to upload your sketch.

With the Arduino now set up to send a message to Processing, you need to set up the Processing sketch to receive that message over the serial port.

Setting up the Processing code

The Processing code is found within multiline comments markers (`/* */`) at the bottom of the Arduino Graph sketch. Copy the code within the comment markers and then paste it into a new Processing sketch, saved with an appropriate name, such as Graph.

```
// Graphing sketch


// This program takes ASCII-encoded strings from the serial port at 9600 baud
// and graphs them. It expects values in the range 0 to 1023, followed by a
// newline, or newline and carriage return

// Created 20 Apr 2005
// Updated 18 Jan 2008
```

```
// by Tom Igoe
// This example code is in the public domain.

import processing.serial.*;

Serial myPort;        // The serial port
int xPos = 1;         // horizontal position of the graph
float inByte = 0;

void setup () {
  // set the window size:
  size(400, 300);

// List all the available serial ports
// if using Processing 2.1 or later, use Serial.printArray()
println(Serial.list());  // I know that the first port in the serial list on
  // my mac is always my  Arduino, so I open Serial.list()[0].
  // Open whatever port is the one you're using.
  myPort = new Serial(this, Serial.list()[0], 9600);

  // don't generate a serialEvent() unless you get a newline character:
  myPort.bufferUntil('\n');
  // set initial background:
  background(0);
}

void draw () {
  // draw the line:
  stroke(127, 34, 255);
  line(xPos, height, xPos, height - inByte);

  // at the edge of the screen, go back to the beginning:
  if (xPos >= width) {
    xPos = 0;
    background(0);
  } else {
    // increment the horizontal position:
    xPos++;
  }
}

void serialEvent (Serial myPort) {
  // get the ASCII string:
  String inString = myPort.readStringUntil('\n');
```

```
    if (inString != null) {
      // trim off any whitespace:
      inString = trim(inString);
      // convert to an int and map to the screen height:
      inByte = float(inString);
      println(inByte);
      inByte = map(inByte, 0, 1023, 0, height);
    }
  }
```

Click the Run button to execute the Processing sketch, and an applet appears. The applet has a black background with a purple graph representing the analog Arduino input, as shown in Figure 14-6. As you turn the potentiometer, the purple graph changes to match it. The graph is updated over time, so as the reading progresses, the graph fills the horizontal space. When it reaches the edge of the display window, the graph resets to the starting point, on the left side.

FIGURE 14-6: A purple graph showing your sensor reading.

If you don't see a graph, double-check your wiring:

- » Make sure you're using the correct pin number.
- » Make sure that the potentiometer is wired the correct way around.
- » Check that you're using the correct serial port.
- » Check that your Arduino code uploaded correctly and that your Processing code has no errors. Note that you cannot upload while the Processing sketch is communicating with your Arduino, so stop the sketch before uploading.

Understanding the Arduino Graph sketch

In `setup`, the code just needs to initialize the serial port. The serial port is set to a baud rate of 9600, which must match the baud rate in the Processing sketch.

The analog input pins are set to input by default, so you don't need to set their `pinMode`:

```
void setup() {
  // initialize the serial communication:
  Serial.begin(9600);
}
```

In `loop`, a single line prints the value of the sensor to the serial port. The pin is directly named rather than being given a variable (such as `analogPin`) because no repeat mentions occur. This pin is `A0`, or analog input pin 0:

```
void loop() {
  // send the value of analog input 0:
  Serial.println(analogRead(A0));
```

Analog readings are made extremely quickly, usually more quickly than they can be converted to digital format. Sometimes this speed causes errors, so a short delay of 2 milliseconds between readings can help stabilize the results. Think of it like a tap that limits the flow of water:

```
  // wait a bit for the analog-to-digital converter
  // to stabilize after the last reading:
  delay(2);
}
```

Understanding the Processing Graph sketch

When data is sent to the serial port, Processing reads that data and interprets it to draw the bar graph. First you import the serial library into the sketch and create a new instance of it. In this case, the new serial port object is called `myPort`:

```
import processing.serial.*;

Serial myPort;        // The serial port
```

One integer, defined as xPos , keeps track of where the latest bar in the bar graph is drawn (the x position):

```
int xPos = 1;         // horizontal position of the graph
```

In setup, the display window is defined as 400 pixels wide and 300 pixels tall:

```
void setup () {
  // set the window size:
  size(400,300);
```

To find the correct serial port, Serial.list is called and printed to the console with println. The function println is similar to Serial.println in Arduino but is used in Processing for monitoring values. These values print to the console rather than the serial port and are used for debugging rather than communication:

```
  // List all the available serial ports
  println(Serial.list());
```

Your Arduino is likely to appear at the top of the list, so myPort uses position 0 as the Arduino serial port. If you are not using the Serial.list function, you can replace Serial.list()[0] with another number in the list, which prints to the console. You can also replace Serial.list()[0] with the name of the port, such as /dev/tty.usbmodem26221 or COM5. Specifying the name is also useful if you have multiple Arduinos connected to the same computer. The number 9600 refers to the baud rate, the rate at which you are communicating with the Arduino. If the rate is not the same on both ends, the data will not be received:

```
myPort = new Serial(this, Serial.list()[0], 9600);
```

In this example, you have another way to sort out the good data from the bad. The serialEvent function triggers every time data arrives in the serial buffer. This line checks to see whether a character followed by a newline character is sent, such as Serial.println(100) on the Arduino side. Newline, or Return on a keyboard, has an ASCII character, which is referred to as \n. The line could also look for other special characters, such as a tab, which is \t.:

```
  // don't generate a serialEvent() unless you get a newline character:
  myPort.bufferUntil('\n');
```

To start, the background is colored black:

```
  // set initial background:
  background(0);
}
```

```
void draw () {
  // everything happens in the serialEvent()
}
```

This bar graph is detailed, with one bar represented by a column of pixels. The `line` function creates these awesome visuals. To change the color of the line to purple, the `stroke` value is set to its RGB components.

The line is defined by its start and end point. Because we're displaying a graph, you want lots of vertical lines with varying heights. As you can see, the x, or horizontal, coordinates are the same variable. The height coordinate for one end of the line is equal to the height of the window, which fixes the height coordinate to the bottom edge of the window. The other height value is equal to the height minus the `inByte` value, so the greater the value, the nearer the top of the line is to the top of the window:

```
void draw () {
    // draw the line:
    stroke(127, 34, 255);
    line(xPos, height, xPos, height - inByte);
```

TIP

If you're having trouble selecting a color, choose Tools➪Color Selector on the Processing menu. The Color Selector displays a color wheel with the red, green, and blue values — as well as the hexadecimal value — of any color you select, as you can see in Figure 14-7.

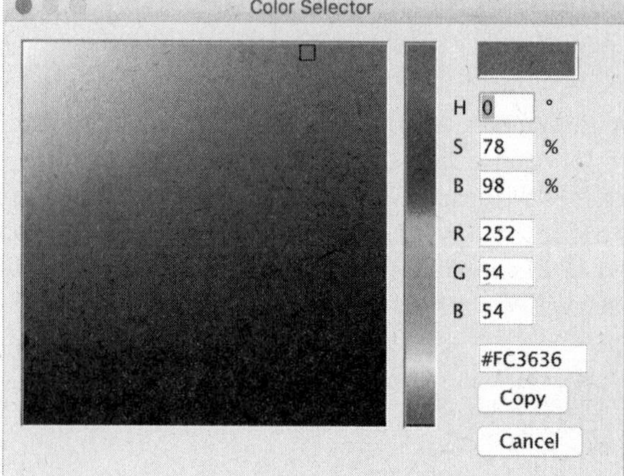

FIGURE 14-7: The built-in color wheel can be useful.

This next bit of code handles the movement of the `xPos`, or horizontal, position of the graph over time. If `xPos` is greater than or equal to the width, the line reaches beyond the edge of the display window. When this happens, the variable is returned to 0, and a new background is drawn to cover the old graph:

```
// at the edge of the screen, go back to the beginning:
  if (xPos >= width) {
    xPos = 0;
    background(0);
  }
```

If xPos is not equal to the width, it is increased by one pixel for the next reading:

```
  else {
    // increment the horizontal position:
    xPos++;
  }
}
```

The serialEvent function is part of the serial library and triggers when data arrives in the serial buffer. Because the bufferUntil('\n') condition has been used, serialEvent triggers when a newline character is buffered:

```
void serialEvent (Serial myPort) {
```

A temporary string is declared to store the data read from myPort. Data is read until a newline character appears. Because the Arduino is sending an integer followed by a newline followed by an integer and so on, each value is read individually:

```
// get the ASCII string:
  String inString = myPort.readStringUntil('\n');
```

An if statement checks that the string contains data and is not equal to 0, or null:

```
  if (inString != null) {
```

To make sure that no anomalies exist, the trim function removes spaces, tabs, and carriage returns from the string. The trim function effectively removes all formatting from the characters so that they can be read clearly:

```
    // trim off any whitespace:
    inString = trim(inString);
```

Now the clean string of numbers is converted into a float, called `inByte`. The float is declared on the left side of the equation and is made equal to the float conversion of `inString`. You can also use parentheses around the variable to convert it to other types of data, such as `int()` or `byte()`:

```
// convert to an int and map to the screen height:
inByte = float(inString);
println(inByte);
```

The newly declared `inByte` is then mapped, or scaled, to a more useful range. The range of the sensor is 0 to 1023, so `inByte` is scaled to a range of 0 to the height of the display window, keeping the display proportional without exceeding the height of the window:

```
inByte = map(inByte, 0, 1023, 0, height);
```

This example is a great exercise for getting familiar with communication between Arduino and Processing. The Arduino could easily be sending data from an analog sensor detecting sound, movement, or light. On the Processing side, things are a little more complicated, but largely because you are generating a complicated visual. The graph could just as easily be an ellipse that gets bigger or smaller as the values change. Why not try it?

Sending Multiple Signals

The only thing better than sending signals to Processing is sending multiple signals, right? Sending values from multiple sensors is easy, but handling them in the correct order on the other end can be difficult. In this example, you learn how to send data from three separate sensors attached to your Arduino to a Processing sketch.

You need the following:

- An Arduino Uno
- A breadboard
- Two 10k ohm potentiometers
- A pushbutton

» A 10k ohm resistor

» Jump wires

The circuit is a combination of three separate inputs. Although they all use the same power and ground, you can think of the inputs individually. Two potentiometers provide two values. These are wired in the same way you would wire a light or temperature sensor, with one side wired to 5V and the other wired to the analog input pin reading it as well as to GND via a resistor. These potentiometers could be replaced with any analog inputs with the appropriate resistors.

The pushbutton provides a digital input as well. One side of the pushbutton is wired to 5V and the other is wired to the digital pin reading it as well as GND via a resistor. Complete the circuit as shown in Figures 14-8 and 14-9.

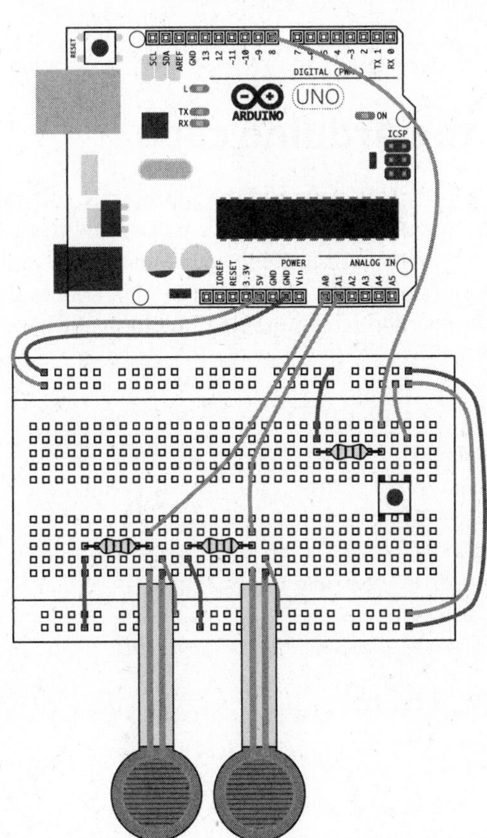

FIGURE 14-8: A circuit diagram for two analog inputs and one digital input.

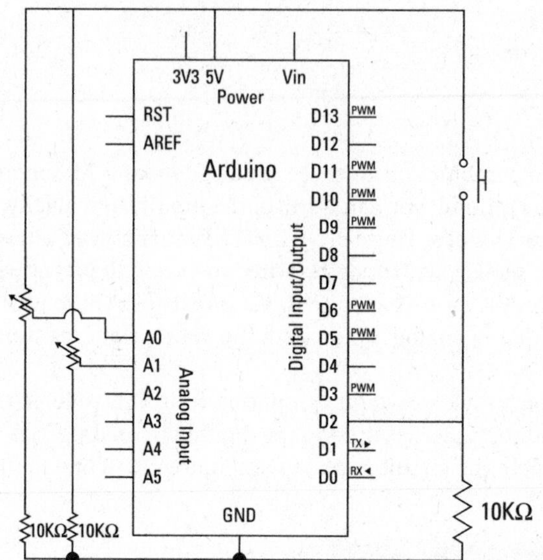

FIGURE 14-9: A schematic for two analog inputs and one digital input.

Setting up the Arduino code

After you assemble your circuit, you need the appropriate software to use it. From the Arduino menu, choose File⇨Examples⇨04.Communication⇨SerialCallResponse. This sketch contains both Arduino code and the relevant Processing code for the sketch to work (along with a variation in Max 5 as well). The Processing code below the Arduino code is commented out to avoid interference with the Arduino sketch:

```
/*
Serial Call and Response
 Language: Wiring/Arduino

 This program sends an ASCII A (byte of value 65) on startup
 and repeats that until it gets some data in.
 Then it waits for a byte in the serial port, and
 sends three sensor values whenever it gets a byte in.

 Thanks to Greg Shakar and Scott Fitzgerald for the improvements

   The circuit:
 * potentiometers attached to analog inputs 0 and 1
 * pushbutton attached to digital I/O 2
```

```
 Created 26 Sept. 2005
 by Tom Igoe
 modified 24 April 2012
 by Tom Igoe and Scott Fitzgerald

 This example code is in the public domain.

 http://www.arduino.cc/en/Tutorial/SerialCallResponse

 */

int firstSensor = 0;    // first analog sensor
int secondSensor = 0;   // second analog sensor
int thirdSensor = 0;    // digital sensor
int inByte = 0;         // incoming serial byte

void setup()
{
  // start serial port at 9600 bps:
  Serial.begin(9600);
  while (!Serial) {
    ; // wait for serial port to connect. Needed for native USB port only
  }
pinMode(2, INPUT);   // digital sensor is on digital pin 2
  establishContact();  // send a byte to establish contact until receiver
                       // responds
}

void loop()
{
  // if we get a valid byte, read analog ins:
  if (Serial.available() > 0) {
    // get incoming byte:
    inByte = Serial.read();
    // read first analog input, divide by 4 to make the range 0-255:
    firstSensor = analogRead(A0)/4;
    // delay 10ms to let the ADC recover:
    delay(10);
    // read second analog input, divide by 4 to make the range 0-255:
    secondSensor = analogRead(1)/4;
    // read  switch, map it to 0 or 255L
```

```
    thirdSensor = map(digitalRead(2), 0, 1, 0, 255);
    // send sensor values:
    Serial.write(firstSensor);
    Serial.write(secondSensor);
    Serial.write(thirdSensor);
  }
}

void establishContact() {
  while (Serial.available() <= 0) {
    Serial.print('A');   // send a capital A
    delay(300);
  }
}
```

Upload this code to your Arduino. Now that the Arduino is set up to send data to Processing, you need to set up the Processing sketch to receive and interpret the data over the serial port.

Setting up the Processing code

You find the Processing code within multiline comment markers (/* */) at the bottom of the Arduino SerialCallResponse sketch. Copy the code within the comment markers into a new Processing sketch and save with an appropriate name, such as SerialCallResponse:

```
// This example code is in the public domain.

import processing.serial.*;

int bgcolor;             // Background color
int fgcolor;             // Fill color
Serial myPort;                       // The serial port
int[] serialInArray = new int[3];    // Where we'll put what we receive
int serialCount = 0;                 // A count of how many bytes we receive
int xpos, ypos;                  // Starting position of the ball
boolean firstContact = false;        // Whether we've heard from the
                                     // microcontroller

void setup() {
  size(256, 256);  // Stage size
  noStroke();      // No border on the next thing drawn
```

```
  // Set the starting position of the ball (middle of the stage)
  xpos = width/2;
  ypos = height/2;
  // Print a list of the serial ports, for debugging purposes:
  println(Serial.list());

  // I know that the first port in the serial list on my mac
  // is always my  FTDI adaptor, so I open Serial.list()[0].
  // On Windows machines, this generally opens COM1.
  // Open whatever port is the one you're using.
  String portName = Serial.list()[0];
  myPort = new Serial(this, portName, 9600);
}

void draw() {
  background(bgcolor);
  fill(fgcolor);
  // Draw the shape
  ellipse(xpos, ypos, 20, 20);
}

void serialEvent(Serial myPort) {
  // read a byte from the serial port:
  int inByte = myPort.read();
  // if this is the first byte received, and it's an A,
  // clear the serial buffer and note that you've
  // had first contact from the microcontroller.
  // Otherwise, add the incoming byte to the array:
  if (firstContact == false) {
    if (inByte == 'A') {
      myPort.clear();          // clear the serial port buffer
      firstContact = true;     // you've had first contact from the
 microcontroller
      myPort.write('A');       // ask for more
    }
  }
  else {
    // Add the latest byte from the serial port to array:
    serialInArray[serialCount] = inByte;
    serialCount++;

    // If we have 3 bytes:
    if (serialCount > 2 ) {
      xpos = serialInArray[0];
      ypos = serialInArray[1];
```

```
      fgcolor = serialInArray[2];
      // print the values (for debugging purposes only):
      println(xpos + "\t" + ypos + "\t" + fgcolor);

      // Send a capital A to request new sensor readings:
      myPort.write('A');
      // Reset serialCount:
      serialCount = 0;
    }
  }
}
```

Click the Run button to execute the Processing sketch, and an applet appears. The applet has a black background, and whenever you press the pushbutton, a white dot appears. Move the potentiometers to move the dot horizontally and vertically. When you release the pushbutton, the dot disappears.

If you don't see the correct behavior, double-check your wiring:

- Make sure that you're using the correct pin numbers.
- Make sure that the potentiometers are wired the correct way around.
- Check that you're using the correct serial port.
- Check that your Arduino code uploaded correctly and that your Processing code has no errors. Note that you cannot upload while the Processing sketch is communicating with your Arduino, so stop the sketch before uploading.

Understanding the Arduino SerialCallResponse sketch

At the start of the sketch, four integer variables are declared. Three are for the sensor values and one stores an incoming byte from the Processing sketch:

```
int firstSensor = 0;     // first analog sensor
int secondSensor = 0;    // second analog sensor
int thirdSensor = 0;     // digital sensor
int inByte = 0;          // incoming serial byte
```

In `setup`, the serial port is established with a baud rate of 9600. The `while` statement continually checks for the presence of a serial connection before proceeding. The statement is interpreted as "while there is not a serial connection, do nothing." This line of code is needed only for the Leonardo boards:

```
void setup()
{
  // start serial port at 9600 bps:
  Serial.begin(9600);
  while (!Serial) {
    ; // wait for serial port to connect. Needed for USB port only
  }
```

Pin 2 is the pushbutton pin and is set as an input by using `pinMode`:

```
  pinMode(2, INPUT);   // digital sensor is on digital pin 2
```

A special custom function called `establishContact` is called to signal to the Processing sketch that the Arduino is ready:

```
  establishContact();  // send a byte to establish contact until receiver
   responds
}
```

In the `loop`, an `if` statement checks whether any data being sent is greater than 0. If it is, that byte is read into the `inByte` variable:

```
void loop()
{
  // if we get a valid byte, read analog ins:
  if (Serial.available() > 0) {
    // get incoming byte:
    inByte = Serial.read();
```

The `firstSensor` variable stores the value of the potentiometer on analog pin 0 after dividing it by 4. This simple math reduces the range 0 to 1023 to the range 0 to 255:

```
    // read first analog input, divide by 4 to make the range 0-255:
    firstSensor = analogRead(A0)/4;
```

A short break of 10 milliseconds occurs to give the analog-to-digital converter plenty of time to process the first value:

```
    // delay 10ms to let the ADC recover:
    delay(10);
```

The second sensor is read and converted in the same way and then the code writes to the secondSensor variable:

```
    // read second analog input, divide by 4 to make the range 0-255:
    secondSensor = analogRead(1)/4;
```

The switch is a little different. To scale the values, the map function is used. The first parameter inside the map function is the number to be mapped. In this case, that number is a digital read of pin 2. Saving the digitalRead value to a variable first in this case is neither necessary nor beneficial. The switch will only return a value of 0 or 1, and it is mapped to the same range as the other sensors, 0 to 255, so it will return 0 or 255 instead. The converted value is stored in thirdSensor:

```
// read  switch, map it to 0 or 255L
    thirdSensor = map(digitalRead(2), 0, 1, 0, 255);
```

Each of the sensor values is then sent one at a time to the serial port by using Serial.write:

```
    // send sensor values:
    Serial.write(firstSensor);
    Serial.write(secondSensor);
    Serial.write(thirdSensor);
  }
}
```

At the end of the sketch is the custom establishContact function, which was called in setup. This function monitors the serial port to see whether a serial connection is available. If not, establishContact sends an uppercase *A* to the serial port every 300 milliseconds. When a connection is available, the function stops:

```
void establishContact() {
  while (Serial.available() <= 0) {
    Serial.print('A');   // send a capital A
    delay(300);
  }
}
```

Understanding the Processing SerialCallResponse sketch

Now examine what's happening on the Processing side to establish contact, interpret the values, and display the data. The first action to take in the Processing sketch is to import the serial library:

```
import processing.serial.*;
```

Many variables need to be declared for this sketch. The first two are colors for the background and shapes:

```
int bgcolor;           // Background color
int fgcolor;           // Fill color
```

A new instance of serial port, called `myPort`, is created:

```
Serial myPort;                     // The serial port
```

An array of integer values is created and declared to be three values long:

```
int[] serialInArray = new int[3];    // Where we'll put what we receive
```

An integer counter is declared to keep track of how many bytes have been read:

```
int serialCount = 0;                 // A count of how many bytes we receive
```

The integer values `xpos` and `ypos` store the coordinates of the dot:

```
int xpos, ypos;                   // Starting position of the ball
```

A Boolean stores a value that indicates whether contact has been made with the Arduino:

```
boolean firstContact = false;        // Whether we've heard from the
   microcontroller
```

In `setup`, the size of the display window is set. The `noStroke` function ensures that no borders are on shapes drawn from this point on in the sketch. Choosing `stroke` or `noStroke` depends on your graphical preferences.

```
void setup() {
  size(256, 256);  // Stage size
  noStroke();      // No border on the next thing drawn
```

The starting values for the dot are set to half the width and half the height, respectively, of the display window, which is the center of the window:

```
  // Set the starting position of the ball (middle of the stage)
  xpos = width/2;
  ypos = height/2;
```

To make the serial connection, a serial list is drawn:

```
// Print a list of the serial ports, for debugging purposes:
println(Serial.list());
```

Your Arduino usually tops the list, so the temporary string `portName` stores the name of position 0 in the list, and `myPort` uses that connection when the serial port is set. Note that this code could be written as `Serial(this, Serial.list()[0], 9600);`.

If you're not using the `Serial.list` function, you can replace `Serial.list()[0]` with another number in the list, which is printed to the console. You can also replace `Serial.list()[0]` with the name of the port. This feature is useful if you have multiple Arduinos connected. The number 9600 refers to the baud rate and must be the same on both ends:

```
String portName = Serial.list()[0];
  myPort = new Serial(this, portName, 9600);
}
```

In `draw`, the background is drawn first. Because the background is undefined, it has a default value of 0, which is black. The fill color is also undefined for now, so the ellipse is black as well. The fill and background colors are updated in the `serialEvent` function, so colored shapes will appear only when there is activity from the sensors.

```
void draw() {
  background(bgcolor);
  fill(fgcolor);
```

The ellipse is drawn, centered on the display window and with a fixed diameter of 20 pixels:

```
  // Draw the shape
  ellipse(xpos, ypos, 20, 20);
}
```

Most of the action happens in `serialEvent`, which affects the ellipse being constantly drawn by the `draw` loop:

```
void serialEvent(Serial myPort) {
```

If data is sent over the serial port, it triggers `serialEvent`. The first byte of data is read by using `myPort.read` and stored in the temporary variable `inByte`:

```
// read a byte from the serial port:
int inByte = myPort.read();
```

If the Processing sketch is not already in contact with a serial device, it proceeds to the next `if` statement to see whether the byte is the character *A*. If so, the buffer clears and the Boolean `firstContact` is set to `true` and sends an *A* character back. Remember that `establishContact` function on the Arduino? It repeatedly sends the character *A* until the connection is made. When an *A* is sent back the other way, it triggers `if (Serial.available() > 0)`, which starts sending the data from the Arduino. This technique is called *handshaking*, a mutual negotiation between two parts of a system to establish a connection:

```
if (firstContact == false) {
    if (inByte == 'A') {
      myPort.clear();           // clear the serial port buffer
      firstContact = true;      // you've had first contact from the
  microcontroller
      myPort.write('A');        // ask for more
    }
  }
```

If `firstContact` is `true`, the program reads bytes as they arrive and adds them in order to the `serialInArray` array:

```
  else {
    // Add the latest byte from the serial port to array:
    serialInArray[serialCount] = inByte;
```

Every time a byte is read, the counter increases by one:

```
    serialCount++;
```

When the counter is greater than 2, all three bytes have been read and assigned their tasks:

```
    // If we have 3 bytes:
    if (serialCount > 2 ) {
```

The potentiometer on analog 0 reflects the horizontal position of the dot, the potentiometer on analog 1 is the vertical position, and the button is the fill color:

```
      xpos = serialInArray[0];
      ypos = serialInArray[1];
      fgcolor = serialInArray[2];
```

These values are also printed to the console for debugging, to check that everything is operating as intended. In Processing, you can combine many values in a `print` or `println` statement by using the + symbol. You can also use `\t` to put a tab in between each value to space them neatly:

```
// print the values (for debugging purposes only):
      println(xpos + "\t" + ypos + "\t" + fgcolor);
```

Another uppercase *A* is sent to trigger the `if (Serial.available() > 0)` conditional and repeat the process:

```
        // Send a capital A to request new sensor readings:
        myPort.write('A');
```

The `serialCount` value is reset to 0 for the next set of values:

```
        // Reset serialCount:
        serialCount = 0;
      }
    }
  }
```

This example is a great way to get started reading multiple sensors into a computer program. Why not build your own giant keyboard to reenact a scene from the Tom Hanks film *Big*? Or gather data from a variety of sensors to monitor the weather around your house? The output for this sketch is basic but offers a huge potential for amazing and creative ways to map data. A few web searches for *data visualization* will give you an idea of what's possible.

5 The Part of Tens

IN THIS PART . . .

Discover websites that will inspire you to learn more about Arduino.

Browse online stores for all sorts of Arduino-compatible components.

IN THIS CHAPTER

» Looking for inspiration on the web

» Searching for tutorials and further reading

» Finding communities in the real world

Chapter 15
Ten Places to Learn More about Arduino

If this is your first step into the world of Arduino, you will be relieved to know that you have an abundance of resources available on the Internet. You can find new Arduino-compatible hardware, projects, tutorials, and even inspiration. In this chapter, I offer ten popular websites to help you on your journey of discovery.

Arduino Blog

`blog.arduino.cc`

The Arduino blog is a great source of all Arduino-related news. You can find news on the latest official hardware and software as well as on other interesting projects. Also found here are talks that the Arduino team wants to share with the community.

Hack a Day

hackaday.com

Hack a Day is an excellent resource for all sorts of technological magic. In addition to presenting a lot of Arduino-related projects and posts, the site offers equal amounts of just about any other category of technology that you can think of. This site contains an excellent collection of posts and information to fuel the imagination.

SparkFun

www.sparkfun.com/news

SparkFun manufactures and sells all sorts of products to make your projects possible, and many of these involve Arduino. SparkFun has an excellent and well-maintained newsfeed that always has some sort of interesting new product or kit to show off. The company also provides excellent videos that explain its kits and document events that the SparkFun team hosts or attends.

MAKE

makezine.com/blog

MAKE is hobbyist magazine that celebrates all kinds of technology. Its blog covers interesting do-it-yourself (DIY) technology and projects for inspiration. Arduino is so important to this community that it has its own subsection in the blog.

Adafruit

blog.adafruit.com

Adafruit is an online shop, repository, and forum for all kinds of kits to help you make your projects work. Its blog announces the ever-growing selection of available Adafruit products as well as other interesting tech news.

Instructables

www.instructables.com

Instructables is a web-based documentation platform that allows people to share their projects and step-by-step instructions. Instructables isn't just about Arduino or even technology, so you can find a whole world of interesting material there.

YouTube

www.youtube.com

YouTube is a great place to kill time, but rather than watching cats do funny things, why not enter *Arduino* in the site's search box to discover new projects that people are sharing. YouTube videos won't always be the most reliable source for well-documented projects, but the videos provide a broad look at Arduino projects in action. Watching videos is especially useful for seeing the proper result of projects.

Hackerspaces

hackerspaces.org

Hackerspaces are physical spaces where artists, designers, makers, hackers, coders, engineers, or anyone else can meet to learn, socialize, and collaborate on projects. Hackerspaces are found in a loose network all over the world. A good place to start to find one near you is the map at hackerspaces.org/wiki/List_of_Hacker_Spaces.

Forum

arduino.cc/forum

The Arduino Forum is a great place to get answers to specific Arduino questions. You often find that other people are working through the same problems that you are, so you're likely to find the answer to almost any problem with some thorough searching.

Friends, Colleagues, and Workshops

Starting out in the world of Arduino can be difficult on your own. You can find many sources on the Internet, but one of the best ways to learn is with friends and colleagues, because learning together teaches you much more than learning on your own can.

Even better is to go to workshops and meet other people. You may find that they have the same interests, allowing you to pool what you know; or they may have different interests, providing an opportunity to show you something new. Arduino workshops are taking place all over the world, so with some careful searching in the Arduino Forum, Hackerspace forums, and Google, you should be able to find a workshop near you.

» Introducing international electronics distributors

» Reusing old parts

Chapter 16
Ten Great Shops to Know

When it comes to buying parts for your project, you'll find a huge and growing number of shops that cater to your needs. Although these stores deal in other hobby electronics as well as Arduino, they stock a variety of boards and components that you can use in your Arduino project. This chapter provides just a small sample of the stores out there, so shop around.

Adafruit

`www.adafruit.com`

MIT engineer Limor "Ladyada" Fried founded Adafruit in 2005. Through its website, the company offers a wealth of resources, including products the company designs and makes itself; other products sourced from all over; tools and equipment to help you construct your projects; and tutorials, forums, and videos covering a wide range of topics. Adafruit is based in New York, New York (it's a wonderful town!). It distributes worldwide and has distributors in many countries.

Arduino Store

store.arduino.cc

Arduino Store was official opened in May 2011 to sell Arduino products directly rather than solely through distributors. The store sells all official Arduino-branded products as well as a few select third-party ones. It also sells TinkerKit, which is designed to make Arduino and electronics even simpler for beginners by using plugs to connect inputs and outputs rather than requiring breadboards or soldering.

Seeed Studio

www.seeedstudio.com

Seeed Studio is based in Shenzhen, China and is self-described as an "open hardware facilitation company." The shop uses local manufacturing to quickly make prototypes and small-scale projects that are distributed worldwide. In addition to manufacturing and selling products, the company offers a community area on its website where people can vote for the projects that they want Seeed Studio to bring to fruition.

SparkFun

www.sparkfun.com

SparkFun sells all sorts of parts for every variety of electronics project. As well as selling Arduino-compatible hardware, it designs and makes a lot of its own boards and kits. SparkFun has an excellent site that acts as a shop front, a support desk, and a classroom for Arduinists. SparkFun also has active (and vocal) commenters on each of its product pages, which help to support and continually improve the company's products. SparkFun was founded in 2003 and is based in Boulder, Colorado.

Allied Electroncs

www.alliedelec.com

Allied Electronics markets itself as "the world's largest distributor of electronics and maintenance products," and is a reliable source of an extensive range of products at low prices. Allied Electronics has a sister company that operates in Europe, RS Components.

Newark Electronics

www.newark.com

Newark is a Chicago-based supplier of electronics with an enormous range of components to choose from. It operates worldwide under the Premier Farnell Group. This company is made up of several sister companies that allow the group to distribute to 24 countries in Europe (Farnell), North America (Newark Electronics) and Asia Pacific (Element14).

Mouser

www.mouser.com

Originally founded by Jerry Mouser, a physics teacher in search of components for a high school electronics program, Mouser Electronics is now a leading US-based electronics distributor. Among its stock is a huge selection of development boards, ideal for those who are just starting out with electronics.

Digi-Key

www.digikey.com

Digi-Key is one of the largest distributors of electronic components in North America. The company originally started by supplying the hobbyist market of amateur radio enthusiasts, but has since grown into an international electronics distributor.

eBay

www.ebay.com

One person's junk is another person's treasure, and eBay is a great source of tech products that need a new home. Many parts are available through eBay — even specific ones. But better than that, you can find other bits of consumer electronics that you can hack to suit your own ends.

Dumpster Diving

Dumpster diving is not technically a shop, but it is a great resource nonetheless! People are always amazed at the amount of useful stuff that's thrown away, but they rarely know what's useful and what's not. The key is to know what you're looking for and find out whether you can salvage the parts you need for your project. This process may take a bit of Googling because so many products and components are available that could be of use to your project.

For example, a motor can be expensive if bought new but is used in a variety of consumer electronics that are often discarded. Printers and scanners use relatively complex and expensive stepper motors, which you could repurpose in an Arduino project. Also, because these everyday objects are mass-produced, even buying a new printer with several motors could be cheaper than buying the motors individually.

Index

Symbols

A

B

C

D

E

F

G

H

I

M

N

Q

R

S

T

U

V

W

X

Y

Z